Wireless LANs

Wireless LANs

Raymond P. Wenig

AP PROFESSIONAL

Boston   San Diego   New York
London   Sydney   Tokyo   Toronto

AP PROFESSIONAL
1300 Boylston Street, Chestnut Hill, MA 02167

An Imprint of ACADEMIC PRESS, INC.
A Division of HARCOURT BRACE & COMPANY

United Kingdom Edition published by
ACADEMIC PRESS LIMITED
24–28 Oval Road, London NW1 7DX

ISBN 0-12-744015-1

Printed in the United States of America
95 96 97 98   IP   9 8 7 6 5 4 3 2 1

Contents

5 Infrared LANs   71

6 Microwave LANs   79

7 Wireless LAN Operations   85

14 The Future Evolution of Wireless LANs   251

Introduction

The day of the fully mobile user is near. Long live the desktop computer and the wired networks. As mobility grows in the workforce, so too will wireless local area networks (LANs.) The wireless LANs will provide the first layer of connectivity between the mobile user and the data and information resources of the world. They will accept interactive requests of the user and move them into the proper servicing level of the system world. Input and outputs will be processed and delivered to and from the mobile users.

Wireless LANs will be a working level of an integrated network world. The user should not know nor care whether their information travels over a wireless or a wired world. The integration of the technologies of wireless and wired will need to be seamless and transparent. The important thing is that the user is able to obtain access to the information resources and services they need to perform their work in a timely manner.

Wireless LANs come in several forms and formats. They can be used to couple a single user to local peripherals and/or a fixed desktop machine. They can couple two small handheld personal digital assistants (PDAs) together. They can also support several roving users in a local confined area. Wireless LANs are usually considered as limited-distance services; however, with proper technology, the wireless LAN can transmit signals over 30 miles with reasonable speeds (1 - 2 Mb/s) which is competitive with a wired T1 line, but for much less cost.

The wireless LAN is not expected to take over the world of communications. However, it adds niche uses and services to the existing network components and extends their uses and performance. When used in appropriate ways, the wireless LAN improves the user's connectivity to the ever-expanding information world. In many applications it will open up new uses and users. In others it will displace some existing network services. The key is to know when, where, why, and how to use wireless LANs within an applications world.

This book concentrates on the use of wireless LAN systems and technology. It provides an overview of the technologies and defines a broad range of uses and applications. The factors for making decisions on wireless LANs, implementation steps, and operational support are given detailed coverage. The interconnection of various networks (wireless and wired) is covered on several levels. Future developments of higher powered hand-held computers, faster wireless protocols, integration to cellular and PCS networks, and other evolutions are also covered.

Wireless LAN networks are likely to be a part of almost everyone's networked universe. From worker to student to executive to private person, the wireless connectivity to other users, business functions, learning resources, emergency support, government support, etc. provides timely, efficient, and effective linkage to useful information and services. Welcome to the wireless LAN world.

CHAPTER 1

The LAN World and the Opportunities for Wireless LANs

Local area networks (LANs) provide services for interconnecting computing resources at the local levels of an organization. LANs can provide services to a small branch office, a department, the floor of a building, a work group, a project team, or any group of users that are within a limited geography. LANs provide connectivity support and services for sharing of resources, passing of transactions, accessing of common databases, messaging between various parties, maintaining the work flow of the organization and interconnecting between the layers of an organization. LANs operate through hardware that makes the physical connections and software that moves the data and transactions between the units.

Traditionally, local area networks have used wire as the medium for physical level interconnectivity. Wire uses copper as the signal medium. Copper wire has been inexpensive, readily available, easy to install, maintenance free, wear resistant, and highly reliable. Wire continues to dominate much of the LAN world today; however, there is a noticeable shift to the use of optics cabling (glass fiber) for specialized high-speed applications.

Wired (and cabled) LANs provide a logical and efficient choice for most networks. Where the users operate from a fixed desk in a standard office environment, the wired (or cabled) LAN is a natural choice. However, if the user is highly mobile and not used to a fixed location of operation, then the wired LAN will be a restriction and a handicap. In addition, some buildings do not lend themselves to being wired or cabled, posing some physical restrictions on the wired (or cabled) approach.

Wireless LANs provide a new layer of flexibility and services to environments and users that cannot be well served by the traditional wired LAN. As more and more users are becoming highly mobile within their building environments, the wireless LAN may be the most effective way to couple the mobile users to their information and communication services. Removing systems tethers from users will enable them to increase their flexibility and productivity to tackle logical business actions and communications wherever they are within the contact space provided by the wireless communication system.

Wireless LANs can be used within almost any building environment. Installation in historic buildings where wire installation would be costly or historically inappropriate can be supported by wireless LANs. Organizations with dangerous or hazardous conditions that could not be supported by wire networks may be amenable to the use of wireless LANs. Emergency situations in which there is no time to lay wire or to establish stable network operating procedures can be supported by an instantaneous setup and implementation of wireless LAN systems. In short, wireless LANs provide LAN world services and internetwork access for a number of unique settings and situations.

This book will concentrate on the wireless end of the LAN spectrum. It will cover the technologies of wireless LANS and explore numerous practical applications for this form of LAN service. The material will provide coverage of both limited distance wireless LANs and longer distance, outside building LANs. Advice will be given on the decisions to be made to select and operate wireless LANs. In addition, the recommended steps to take for a successful wireless LAN project will be described. Some extrapolations on the future of wireless LAN technology and its application will also be provided.

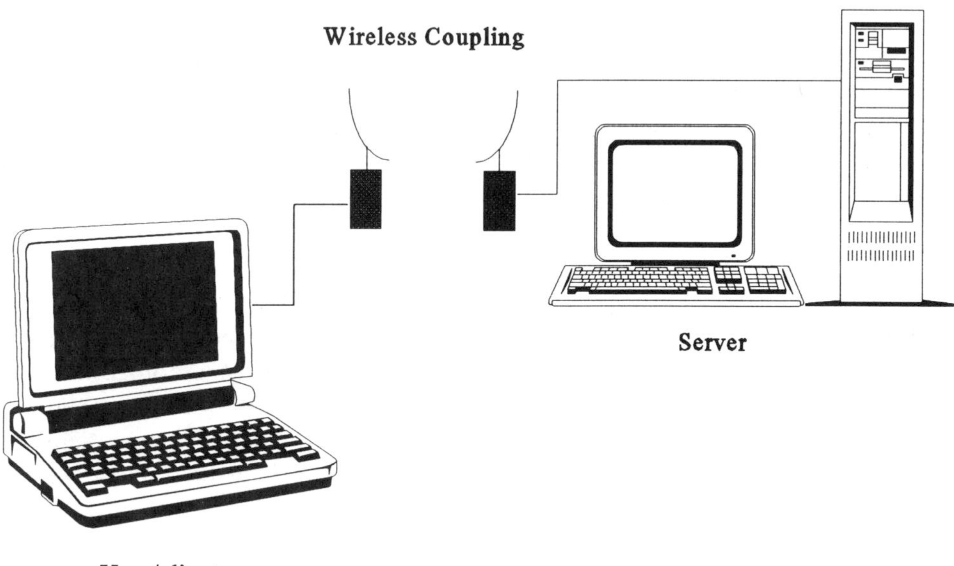

Figure 1.1: Typical Wireless LAN

Wireless LANs are both old and new. The original LAN, the Aloha Net at the University of Hawaii, which the Xerox Corporation used as the basis for its Ethernet proposal, was a wireless LAN. The proposed Ethernet became a wired installation and most of the LANs since have followed suit. New approaches to wireless LANs have developed around cellular services, spread-spectrum radio services, and infrared signaling.

The future of wireless LAN systems will depend on their success in meeting user needs and expectations. The technology will continue to mature and the wireless services will complement the wired world to make a seamless integrated internetwork process. Some future forecasters such as Nickalus Negroponte of the MIT Media Laboratory indicate that "if it moves use wireless, if it doesn't move use wired technology." This may be very simplistic, however it shows the growing importance of the move to support the mobile workforce and flexible operations.

Wireless LANs represent a spectrum of capabilities that support limited distance local coverage, moderate distance metropolitan coverage, and longer distance (anywhere) coverage. Depending on the power of the transmitters and the sensitivity of the receivers, wireless LANs may become the first truly universal form of virtual LAN (VLAN.) By coupling wireless LANs with other wireless communications forms (such as cellular or satellite), the user connectivity can be almost limitless.

Clients and servers

Clients and servers are the hot button for today's application builders. Over 90% of recently surveyed chief information officers indicated that they are or soon will be building client/server applications. The use of distributed PC level clients interacting with distributed servers provides a major opportunity for the wireless LAN. The wireless LAN can provide support for mobile clients and specialty environments where wire media would be impractical, impossible, limiting, or difficult.

A key aspect of client/server systems is the movement of as much of the processing as practical to the client (user) level of the system. The client/server configuration will deliver programs, data, operations, controls, and support from the server levels to the client levels. As clients will be the remote end-user levels, they will be at the end of the network that will be most likely to use the wireless LAN technologies.

As client/server systems focus on user-oriented functions and applications, the portability factor of wireless LANs will appeal to a multitude of users. The ability to reach servers from many points within a local environment will be a major service provided by the wireless LAN. This will allow the portable clients to attach to remote servers and up- and download data and transactions over the wireless LAN.

The wireless LAN will allow mobile workers to maintain their mobility and yet still have full connectivity to the data and resources of their organization. Wireless LANs will become the front end to wired networks and provide the full range of services to the end user.

Client operations

User/client operations will benefit most from wireless LANs. The user/clients will be the principal layer of wireless connectivity. The users will be able to have more locational independence and portability within their operating environment. They will also be able to access a range of LAN services without having to maintain and sustain a physical link connection. It is also easier for a wireless LAN system to access multiple servers or services using various wireless techniques or wireless frequencies.

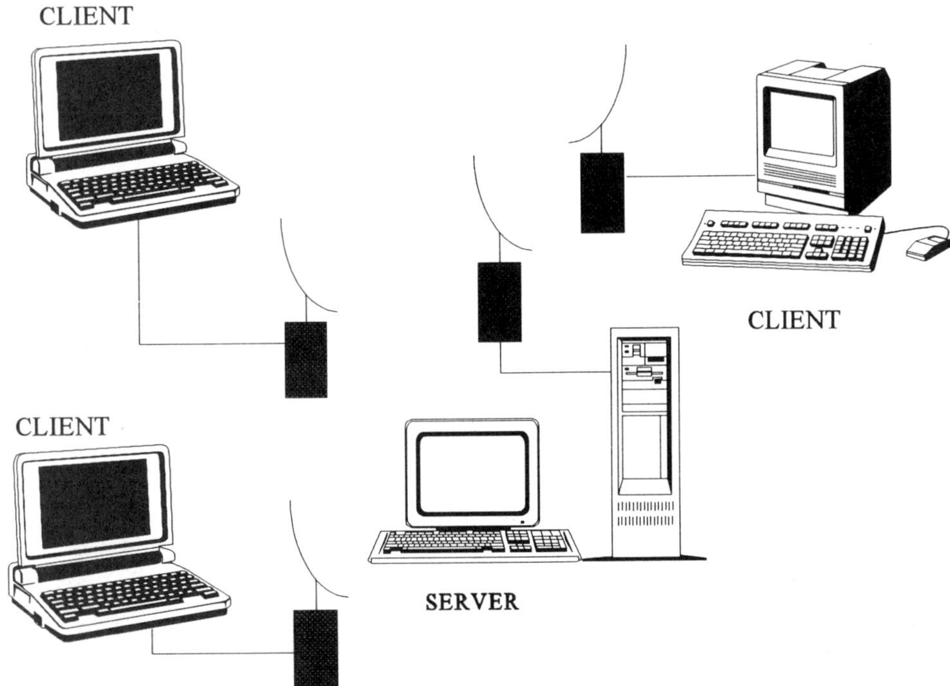

Figure 1.2: Client/server wireless LAN

The operation of wireless clients will provide logical and needed capabilities to several types of applications. People operating on the floor of a stock exchange, moving about in a hospital, operating moving equipment in a factory, moving about the halls of an

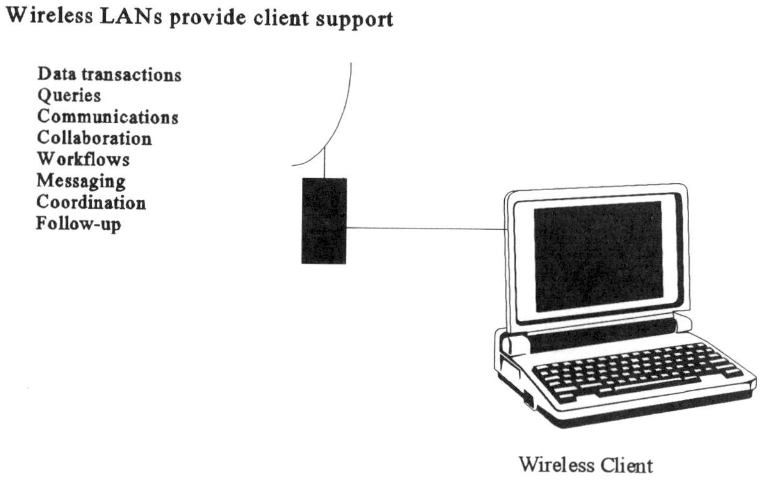

Figure 1.3: Client Support from Wireless LANs

office environment, and in a number of other situations can now be connected to their data worlds without finding and signing onto a fixed terminal or client station. Mobility is the key to wireless clients. It may not be needed by everyone or every application, but it provides a new capability for organizations that can take advantage of it and increase their productivity and performance.

The world above the user/client level may be a mix of wireless and wired systems. This would be transparent to the user/client level, as should the use of a wireless LAN layer. The real underlying issue is the provisioning of reliable, high-availability integrated data and information services.

Wireless LAN services

Wireless LANs will offer the same user services available from wired LANs. The client/user can access files on servers, perform remote operations and applications, transfer data, share resources, and interact with other facilities and users within the enterprise network. Users will be able to communicate, interact, transfer, transact, query, and initiate actions and reactions from within the organization. The major difference of wireless LANs is that they are not connected to wire or cable and users have extensive flexibility to move about and still receive services from their LAN world.It is important that the wireless connections be reasonably transparent to the end user.   Many users will find themselves moving from the wired to the wireless environment and back again. If the systems were different in their operational form, function, and user interfacing, they would not be acceptable in light of the moves to seamless integrated information systems. As the wired LAN world sets new services and standards, the wireless world will need to implement and remain consistent.

In addition, the user will not be able to discern the boundaries of the wireless LAN world. This requires that some form of notice be given when they reach the limits, or the system will have to support some type of roaming to allow the user to move from one area to another and still retain services from the LAN environment.

The differences in wireless LAN services will be found in the slightly slower speeds of information transfer, the problems of security of transmission, and possibly error or noise factors. Many of these differences can be overcome with advances in technology and the evolution of specific wireless components that address these restrictions. For example, by using the available frequency space more aggressively, technology can not only improve the transfer reliability of wireless LAN communications but also improve the security of the information transfer at the same time.

In addition, some of the perceived problems of wireless LAN technology may be patently false. For example, the air waves used by the wireless systems are considered very open and unsecured. However, the spread-spectrum radio format of wireless communications was developed by the U.S. Department of Defense to support battlefield-level secure communications. Other limitations may also be overcome with aggressive reengineering of the technologies.

The security concerns over wireless LANs can also be addressed via the use of encryption techniques and other means of scrambling the contents of the information transferred over the network. Given the ease with which someone can tap or copy the contents moving over copper wire (with minimal chance of detection), the wireless LAN world may provide improved security by forcing the users to take steps to assure the safety and accuracy of the communications over the wireless world.

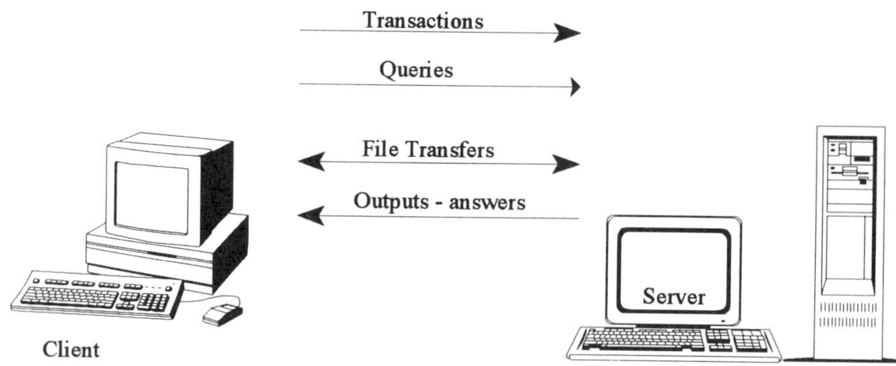

Figure 1.4: Wireless LAN Services

The benefits of the wireless LANs will include flexibility, portability, modest cost, movable installation, and multiple system interconnectivity. The wireless LANs will bring mobile workers into the information world as full-time partners. They will also allow other fixed workers to migrate to mobile operations to improve their performance and productivity within organizations.

Data movement

Movement of data across the wireless LAN will likely be one of its major activities. Input can be scanned or keyed via portable devices and sent to servers for processing and storage by using the wireless transfer technology. Inputs in warehouses, roving data entry specialists, store inventory tracking, health care providers, and a host of other data movement applications will be naturals for the wireless LAN. In addition, by using longer distance wireless LAN couplings, remote users, such as traveling salespersons and route delivery drivers, can be connected to their information and communication systems.

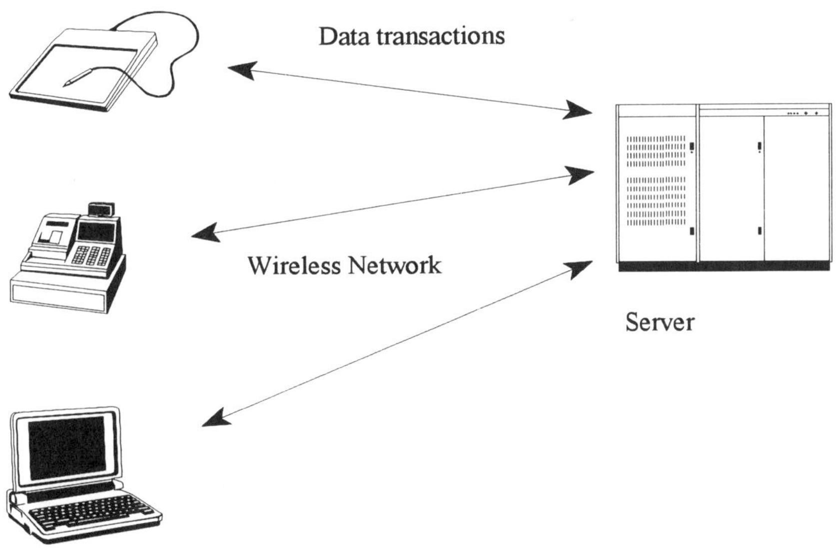

Figure 1.5: Moving Data over Wireless LANs

Data movement will make the wireless LAN a key component in operational systems. It will be the service that justifies the investment in wireless technology and the uniqueness of wireless that allows it to perform services that no other technology can duplicate.

Data movement via wireless LANs can be unidirectional or bidirectional. The data being moved can be new data, it can be event recordings or commentaries, it can be extracted data moving from servers to the user, or it can be combinatorial data assembled from several sources. It can be text, image, animation, black and white, or colored, it can be short or long, it can be detailed or summarized, and it can contain data or descriptions. In short, the wireless LAN can handle any type of data that exists in an information systems environment.

The early wireless communication systems were built primarily to carry data between various type of instruments. The early systems moved data between technical instruments, scientific controllers, and medical instruments. These services were named ISM in honor of the instruments, scientific, and medical applications they supported.

Interoperability

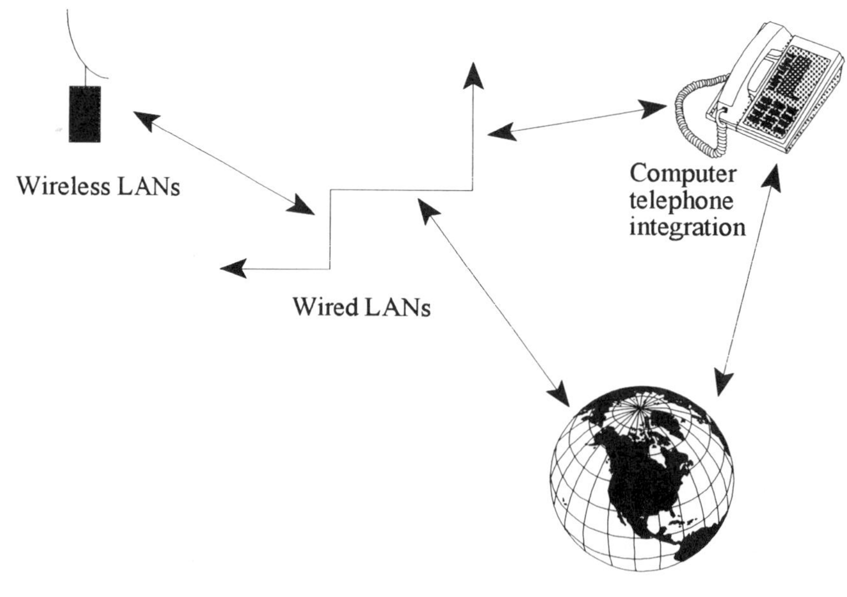

Figure 1.6: Integrated LAN Worlds

The wireless LAN will become an integral part of operational enterprise networks and services. This will make it an interoperable part of the total information processing system. It will also provide users on the wireless LAN with full features and functions available to any other user on an interconnected network.

The interoperability on the wireless LAN may be limited for user and/or security reasons; however, this is a design decision and not a functional limit of the technology. A wireless LAN node should be able to access and use any service available to its user community and interconnect to any servers or services that would be logically available to its operation or business function.

The issue of security in wireless LANs can also be dealt with through the use of frequency-hopping services or data encryption. With the proper application of technical design options and use of normal security techniques like user passwords, wireless LANs can be made just as secure as wired LANs.

Wireless LAN support requirements

Wireless LANs will require some unique support and service functions. The facilities in which they operate will need to be free of major wireless interference and the transmitters and receivers will need to be mounted and tuned to the operational environment. Wireless LANs will also need monitoring and control to maintain reliable and secure services and provide some mechanism for performance management and control.

Wireless LAN clients, often being portable and mobile, will require some form of physical security and validation before being allowed open access to the LAN. Version control and status validation may also be needed for mission- critical applications. In applications where the processing is under tight audit control, the support process may need to download authorized software each time the wireless unit is connected to assure proper operational components.

Wireless LANs will also require more remote types of support. If the unit breaks down at a mobile location, the service or replacement service must be able to respond quickly and efficiently to the location of the failure. Once the failure is detected, repairs must be made and the remote service reestablished with the users. This places a geographic coverage requirement on the support services, which could be large and expensive to cover with properly qualified talent. The alternative to remote support is to provide users with highly reliable units which may have some redundant fail-soft features built in that can be automatically switched into operation in the event of a fault. In addition, alternate routing services may be built into the wireless network to allow the users to receive services from backup facilities that are transparent to the individual user.

The more support that can be built into the portable wireless LAN equipment and the network components, the better will be the reliability of the service and the lower the costs of personal support. Even if the user has to be equipped with some computer cards, Personal Computer Memory Card International Association (PCMCIA) units, a spare transmitter, etc., these costs should be less than those of providing a cadre of qualified people who can appear at any location to repair a fault when it occurs.

User activities

The users of wireless LANs should see little or no difference in their operations. Their client stations provide them with full facilities of an independent unit and the wireless LAN network will function almost the same as if it were a traditional wired LAN. Access delays, contention, errors, and other problems should be tunable to a level that is the same as or maybe better than those experienced in a wired environment.

Users will see some differences in the physical units they will be using to interact with their information world. The large, fixed, desktop computer will be used only in a small number of wireless LAN situations. Most of the user interface devices will be small, personal, and portable devices that can go anywhere and operate from anyplace to become the users client level processor. The user client devices can range from notebook computers to personal digital assistants (PDAs) to speciality devices such as handheld bar code readers. The variety of client devices will vary with the range of applications and situations.

The major improvement for user activities with the wireless LAN units is that they will be able to operate in more flexible and mobile ways within their environment. There will be range limits for connectivity, and there may be blackout areas where the wireless LAN is unavailable. These are similar to limits on wired LANs; however, the mobile user may experience these limits during the usage process where the wired system is more stable and reliable in its connectivity services. The loads on wireless LANs may be more predictable and continuous than those on wired systems, providing a more consistent availability of services. In addition, by limiting the number of users on a wireless LAN, the services and throughput may end up being better than those in the wired environment.

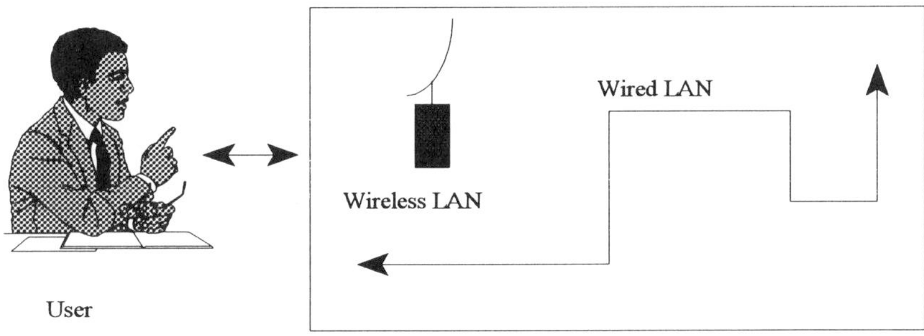

Figure 1.7: User Transparency

Wireless LAN advantages

There are numerous advantages to the use of wireless LANs. They include:

- Flexibility to support previously excluded applications and users

- Ease of installation

- Ease of modification

- Mobility

- Portability

- Interconnectable

- Expandable

- Easily segmented

- Economics

- Transparency

- Reusability

- Broad range of coverage and options

The wireless LAN is not intended for all applications and situations. The technology is stable and cost effective and it is appropriate for selected applications. Where the technology is appropriate, the wireless LANs are a most logical choice. When the situation is marginal, other choices may be more effective.

The advantage of covering previously unsupported applications with wireless LANs allows for the opening up of numerous areas, applications, and users to becoming full partners in the information services worlds. These people had previously been denied access to this world or had to make access through inconvenient means and techniques, thus limiting the effectiveness of the connections. With the wireless LANs, these users can access almost any layer and element of information processing while still maintaining their mobility and flexibility within the organization.

The second advantage of wireless LANs is the ability to initiate a fast setup or a quick move of the technology and continue to offer connectivity services. This is critical for emergency services and users who must relocate themselves while still needing access to organizational information. The flexibility of wireless LANs to be installed quickly and come up running with minimal setup and initiation makes them a logical choice for most emergency or temporary settings. Whether at a fire or a wedding or a sports event, the wireless LAN can supply data transfer services at reasonable speeds with minimal setup time and effort.

Wireless LANs should be considered as having two layers of capability. One is the limited coverage of a local area, where the wireless capabilities compete with those of wired LAN technology. The second layer is the remote services where the distances are beyond the coverage space of any wired technology. By using microwave transceivers or portable satellite dishes, the wireless LAN distance can be almost limitless. In fact, the longer distance wireless LANs will end up competing with traditional communications services such as data communications lines, frame relay, and other telecommunications services. The advantage is that the wireless LAN services will be user and application dedicated and may use less costly communications components that those supplied by the traditional telecommunications firms. As an example, a satellite office could be coupled with a long-distance wireless LAN and then take advantage of all of the services provided at its primary business location. The remote users could exchange data, tap into the corporate PBX system, use remote databases, communicate to corporate personnel, and maintain apparent contact as if they were a physical part of the central organization rather than a small satellite facility.

The longer distance wireless LANs may be very appropriate for telecommuting and other remote user operations. If the costs of the wireless services are competitive with those of alternative communications, then the flexibility, reusability, and scalability factors will make the wireless form the best way to go.

Figure 1.8: User Flexibility from Wireless LANs

Wireless LAN flaws

The flaws of wireless LANs are found mostly in their environmental limits of connectivity and in their reliability of operation in all areas and at all times. At the edge zones, the connections between user clients and the network can become error prone, easily broken, and/or distorted. As long as the wireless mobility is kept within tight limits of the capabilities of the installation, the flaws of wireless LANs should be equivalent to those of wired units.

The biggest flaw for wireless LANs is when they frustrate the end user. The user wants reliable and effective services from all of their computing world. If the wireless LAN is unable to deliver reasonable services when the user demands them, then the product will be avoided or rejected. There is little patience or sympathy for an un-reliable and non-performing technology today. Uptime and service are the watchwords for all systems. Quality, performance, and above all, user service is the only survivable game.

Wireless LAN frustrations

    Lack of reliability
    Errors
    Long waits
    Incomplete returns
    Contention
    Slow response
    Unpredictable timings
    Instability
    Variable space coverage

Frustrated user

Figure 1.9: Wireless User Frustrations

As a new technology, wireless LANs still have teething problems. They can operate erratically and are subject to many forms of interference. If these flaws are not overcome, then these products will be unsuccessful in the marketplace. Even if these problems are not inherent in the wireless LAN systems, the products will still have to prove themselves in test environments and with "bleeding edge" users. Only after confirmation and proof of the technology and its stability, will the markets openly welcome this technology to its applications menu.

Most IT users are still from "Missouri" as witnessed by their constant cries of "show me—prove it to me that this technology works."

The other key flaw is in the limited speed and transfer rates of the wireless systems. The limits of 1 or 2 million bits per second (Mb/s) could hamper the use of wireless as a high-speed file transfer service. Plans for 10 to 16 Mb/s transfer speeds would put the wireless technology in the same range as most current LANs. However, as the wired technology moves up to the 100 Mb/s range, it is unlikely that the wireless systems will be able to reach this new high speed plateau. Speed, however, may not be that much of a limiting factor, especially in situations where the client is a small (and limited capacity) laptop or notebook computer and the transferred data are small transactions with limited movement of large data blocks and files.

The low-speed individual wireless links can also be partially overcome by the use of dedicated channels and frequencies for each user (similar to hub switching on wired LANs). This would provide dedicated private line couplings from user to the server worlds where the 1 to 2 Mb/s speeds would be higher than the shared services rates on party-line wired LANs. The real key lies in the ability of wireless technology to move information reliably and accurately between the users and the service facility. If users are satisfied by remote coupling via a 9600 baud modem, then they will be ecstatic with a 2 Mb/s wireless LAN connection.

In short, the current flaws of wired LANs can be dealt with via good design and layout. Technology will be able to offer several long-term solutions to overcome most of the fundamental restrictions of the current systems.

Wireless versus fixed wiring media

Wireless LANs will never replace fixed wiring media. This was never the intention of this technology. However, there are several applications that are not well served by wired media. These applications may be served by the wireless systems. In the end, the wireless and the wired media systems will be integrated at some level. The higher levels of enterprise networks will continue to be primarily built of wired media. The wireless LANs will operate at the lower levels and interface to the wired networks for the higher levels of service.

The issue should not be wireless versus wired in local area networks, but rather when one should use which technology and where they interconnect. Eventually, the complete spectrum of LAN options and alternatives will be available for any application. The wired versus wireless argument will be one of adaptivity rather than competition.

Economics of wireless LANs

Costs and benefits will still be a part of the consideration of the use of wireless LAN technology. The good news is that the air medium of wireless systems is less costly than the copper or fiber medium of the alternative technologies. However, the transmitter and receiver products are still somewhat more expensive than the same functional units in the wired world. This is primarily due to the popularity and volume of the wired products and the novelty of the wireless units. As the popularity and acceptance of wireless technologies grow, the prices of the transceivers will drop and the economics will shift to favor the wireless world. In addition, if the vendors begin to build wireless capabilities into their basic system units, the costs will drop even further. Consider now the large number of notebook computers that are coming with builtin infrared wireless capabilities. Stay tuned; as the popularity of wireless grows, the volume will build and the prices will drop. This will in turn change the economics of the decision and tilt it in favor of wireless. Given the other advantages to be derived from wireless LANs, the current extra costs can often be overcome with reasonable valuation of the other additive benefits such as flexibility, speed of installation, and scalability.

CHAPTER 2

Wireless LAN Architectures and Choices

Wireless LANs are a range of technology, not a singular implementation. The variant technologies provide different services and capabilities and operate in vastly different ways. The key feature of the wireless LANs is that they do not require a fixed wire connection. The wireless connectivity is supplied by framing and transmitting signals over frequency space that does not require physical wire containment.

The open air space surrounding the user's environment is the major host for the wireless communications. This means that available forms and frequencies within the transmission spectrum will be used for the wireless LANs. This frequency space is already chopped up and assigned to various uses and users. The wireless LANs need to fit within frequencies that can be reused, assigned within current usage settings, or used at such low power that they do not affect other users.

The major alternative technologies will be examined and evaluated in this chapter. The component parts of wireless LANs will be defined and explained. The software and procedures that bond the wireless LANs to the information systems and services will also be described and positioned in relationship to the overall architecture of wireless LANs. Both short- and long-distance wireless network components will be examined.

What's different

The wireless LAN interconnects the users to an information services world using some form of flexible, nonwire connection medium. The services offered to the user/client will be the same as those available through a standard wired LAN or close thereto. What's different is the flexibility and portability afforded to the user. Users are no longer tied to a fixed tether. Wireless LANs can be mobile (within limits of geography), they can operate with portable client computers, they can be moved and reassembled quickly without the hassle of laying and connecting wire.

The services offered through a wireless LAN will be a replication of those provided through a wired LAN. The network operating system will be the same or similar, and the commands and actions will be the same. The differences in the user's view will be minimal. The differences from an installer and an operational support view will be more dramatic and significant. But these differences will be in the background and remote from the user. The differences will occur only at installation or change times, so will be nearly transparent most of the time for the operations level.

Why wireless?

Wire has been the chosen medium for LANs since they were invented in the early 1970s. Why have the need for and interest in wireless developed over the past few years? The answer lies in the fact that some applications have found it difficult or impossible to implement LANs. For example, consider a warehouse, where there is a need to track vehicles, movements of goods, locations, etc. in support of a logistics and distribution activity. Wiring LANs for such an environment would be difficult to impossible. Yet with wireless LANs, the roving users can be in direct tw- way contact with the full resources of the enterprise's information systems. Similarly, situations in which people move around to various locations to perform their work (as opposed to staying at a fixed desk that can be easily wired) are candidates for wireless LANs. Such situations could include manufacturing stations, assembly lines, researchers, testers, health care providers, and a host of others.

The wireless LAN opens up the coverage of LAN services to a host of applications that have been isolated from the wired world. The answer to "why wireless" is that it fulfills a need that has heretofore gone unanswered. The wireless LAN is also opening up new applications that had not even considered the possibility of being part of a LAN. In addition, several previously wired applications are now considering the potential of going to wireless.

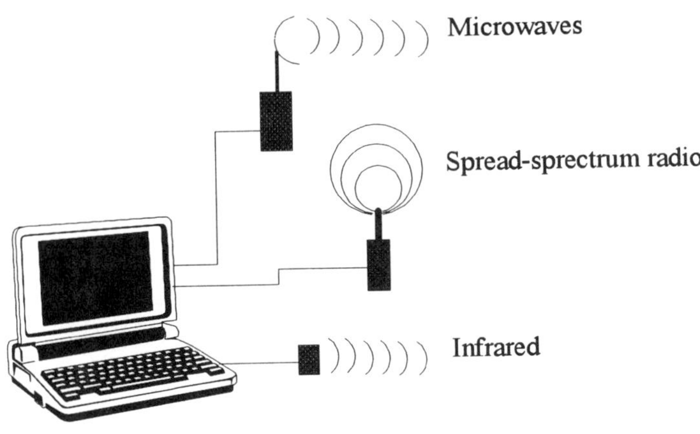

Figure 2.1: Differences of Wireless LANs

Wireless LANs can also move to higher power plateaus which will extend their distances from hundreds of feet to several tens of thousands of feet. In fact, with the multimile capability of some of the wireless networks, this technology may be more far ranging than the wired world. Add in wireless satellite broadcasting, and you have a global capability. These alternatives can make the wireless LAN world more flexible and support longer distances than the other options.

The wireless LAN components

Wireless LANs have two components additional to or different from those found in a wired LAN. These components are the waveform transmitter and the receiver that handles the data transfers from the user client and the servers to the air space. The transmitters and receivers can use a variety of media and frequencies to support their services. The wireless LAN will still need all of the components that are used in a wired LAN, including a network interface card, a network operating system, application programming interfaces, network-compliant software, and various operational procedures.

The waveform transmitters and receivers for the wireless LANs will vary with the specific technology and wave frequency used. The radio frequency wireless LAN is different in level, form, power, and other factors from the infrared units or the microwave stations. These differences will result in different power requirements, different antenna systems, different tuning and operational support requirements, and different performance values. The differences will also cause variations in user operations, transparency, ease of use, and scalability.

The waveform transmitters will convert the signals of the computer system into waveform motions at the correct frequencies and send them into the air medium at the proper time and signal level. The waveform receivers will extract the signals from the air and perform review, recognition, and acceptance processing on them. The transmitters and receivers will require external antennas to enter and receive the waveforms from the air medium.In addition to the different components of the wireless LAN configuration, there is the added requirement of careful design and setup of the equipment and the tuning of its sending and receiving operations. Issues such as facility obstructions, antenna locations, signal tuning, electromagnetic radiation testing, and movement patterns must be evaluated and taken into account before committing to a wireless LAN installation. There may be conditions and situations that would limit the usability of the wireless LAN technology and restrict the success of the investment. It is also possible that one wireless LAN technology may work better than another in given situations. A trial setup or a pilot installation may be needed to make the final verification of wireless LAN feasibility.

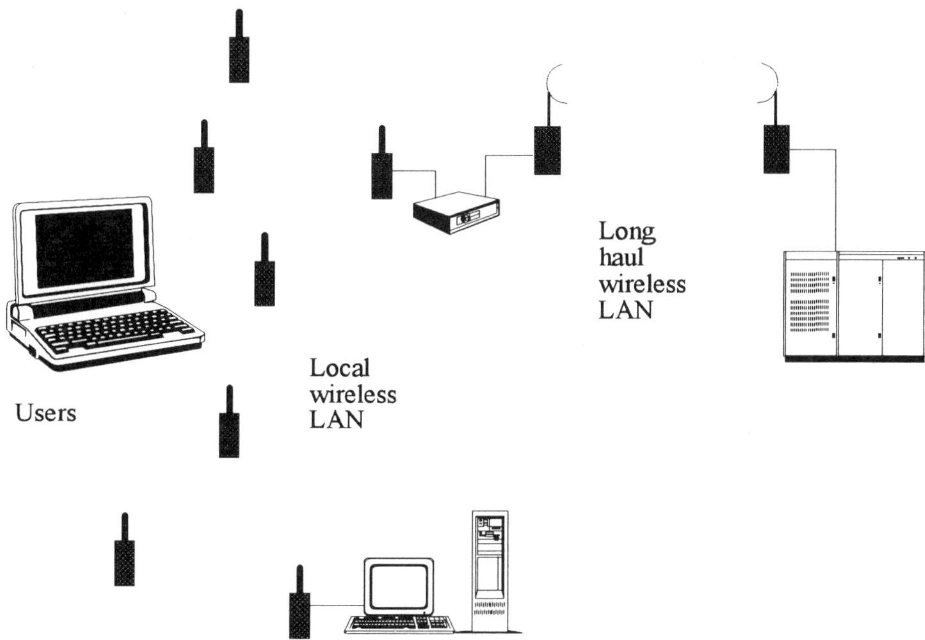

Figure 2.2: Wireless LAN Variations

The use of the air medium

The communication of information between two computer systems requires some form of connecting medium. The wireless LAN uses air as its medium rather than a physical wire. Air is a part of the fluid world and behaves according to defined laws of physics. The principal physics for communication is the pulsing of the medium in such a way that the pulses travel across the medium and can be detected and reliably received at specified points in the medium. The medium must carry the signal with accuracy, reliability, continuity, and quality.

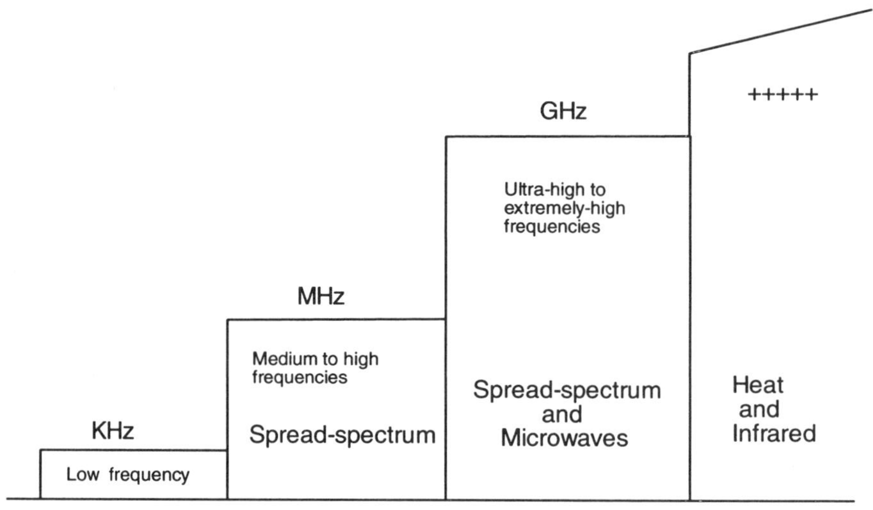

Figure 2.3: Waveform Variations

The physics of the use of the air medium is actually the same as that used in the wired world. Pulses are generated at one connection of the wire, and they travel through the medium (wire) and are received (detected) at other points (stations) along the wire. The difference between the medium (wire versus air) only dictates that the means of generating the pulses may be different. In a wire medium the pulse is set and maintained by electrical current turned off at specified intervals and maintained at a set power level.

The air medium is also pulsed, but with waves that can be transmitted by broadcasting into the medium using predefined frequencies. The available frequencies are defined by wave mechanics and physics and the spectrum of frequencies allocated by laws and regulation for various purposes.

Three basic levels of pulsing are used for today's wireless LANs. One is the radio frequency band space, another is in the very small wavelength area known as microwaves, and another is in the frequency area just above visible light, which is known as infrared. Several other frequencies are available in the air medium pulsing arena, such as low-frequency heat variations, frequencies just out of the human audibility/hearing range, and several others.

The use of the air medium requires that wireless LANs have a transmitting and a receiving capability to pass high-reliability, high-speed, error-free messages and control information between one another within their prescribed (or limited) geographic area. The *local* of local area networks will have precise meaning in terms of distance and power and accuracy of the wireless system.

The air medium is not as "clean" and reliable as a contained wire medium. The wire environment is dedicated to the communication process and is a carefully defined physical world. The air is common and shared by many uses and users. Air is also not a particularly clean world. Continuous and dynamic disturbances are placed on the air which can cause disruptions and problems in its use as a communications transfer medium. The movement of heavy physical objects through the air causes the air to move in different patterns and waves. Thus vehicles, people, machines, etc. can move the air medium in different ways and induce distortions in the flow of the communication waves through the air medium.

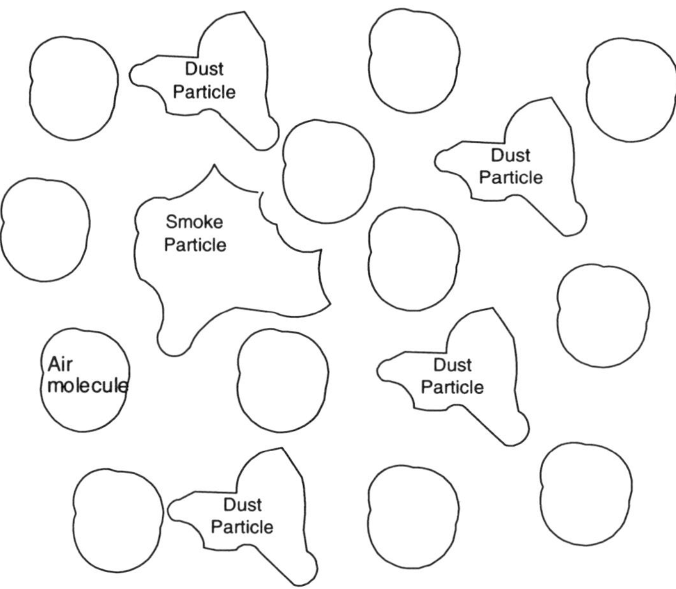

Figure 2.4: Air Signal Movement Problems

Basically, air is an unstable, multiuse, multiuser environment that is not particularly friendly to the hosting of communications signals. The air can be contaminated with foreign particles, distorted by operations and movements, and disrupted by influences such as electrical interference, lightning, and other phenomena. In some respects it is amazing that air can be used at all for the reliable transmission of communications signals.

The physics of the air molecules is also different from that of the wired world. Air is composed of loosely coupled molecules containing a variety of different components. To pulse these molecules requires more power and patience, as the separation of the air components is greater than that of tightly packed copper atoms.

Coupling options

The coupling of wireless LANs involves connecting the computing units (clients and the servers) to the wireless medium. This is usually done with a transceiver on each unit that is tuned to the specified frequency of the communications process. The transceivers form a combined unit that is capable of both transmitting and receiving the signals for the wireless medium. The units will also have to be capable of formatting and recognizing the transmission packages that are being sent over the wireless medium.

The coupling of units for wireless LANs can be done at many levels. The simplest is a one-to-one coupling between a portable client and a fixed server system. The next level is one-to-many, where a portable client can have wireless connectivity to a series of servers. The most extensive level is many-to-many, where several portable users can interface to many remote servers and services.

The wireless LAN transceivers will have to convert the network data packets from the internal computer format (probably Ethernet standards) to the format defined by the transceiver vendor. The standards at the transceiver level are limited, with most of the wireless LANs depending on the unique format defined by the producing vendor. At the receiving end, the vendor's transceiver reconverts the data packets to the standard format for the computer system.

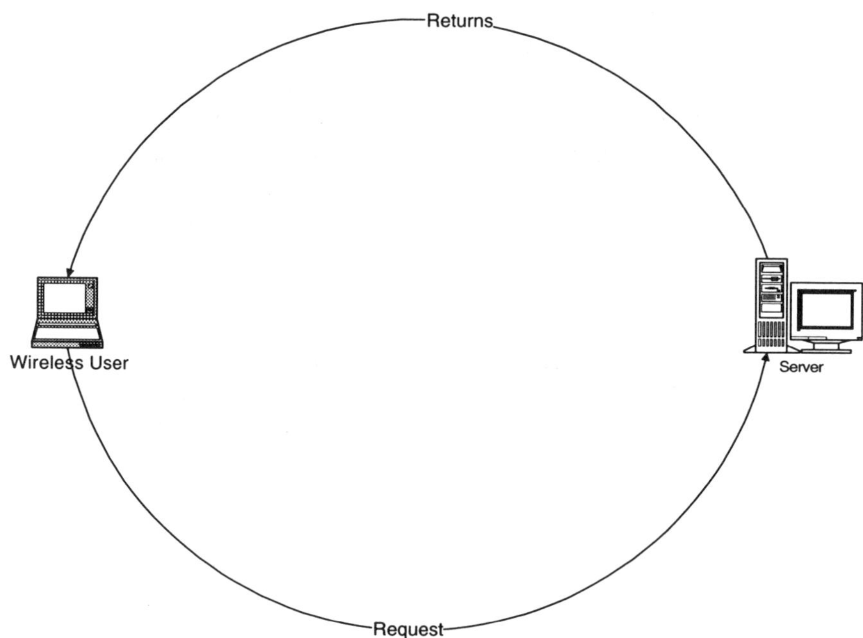

Figure 2.5 Wireless LAN Message Flows

One-to-one operations

One-to-one wireless LAN operations will take place when a single portable unit wants to make data contact with a single fixed server unit. A typical example would be a laptop/notebook computer coupled to a fixed desktop computer station. This would allow for the intertransfer of data, files, programs, and other information between the portable unit and the fixed station. The fixed station can be the connection to a wired LAN world and form a bridge or conduit for passing requested or needed information between the user's client station and the rest of the information world. The one-to-one wireless connection may become the most numerous implementation. The growing number of personal digital assistants, laptop computers, and other portable devices can all use the one-to-one connection.

The one-to-one wireless LAN would cover limited space and traffic. It would be under the direct control of one user, who would most likely control both the portable unit and the desktop system. The simplicity of the one-to-one coupling will make it easy to operate and manage. However, the use of the capabilities of the wireless communications will be limited to the creativity and needs of one individual.

One-to-many operations

When wireless LAN users want to access several different facilities or servers within their networked world, they are coupling to a one-to-many relationship. This connection is more complex than the one-to-one relationship, described above. The user, or the system administration process, will have to know where and how to navigate the network and the attached resources to locate and return the services requested by the user. One-to-many operations will also have to be cognizant of the traffic patterns of other users and applications, determine service facilities and performance factors, control the accuracy and reliability of the operations, and be vigilant for any faults or glitches that could occur while the user is interacting with the other resources.

One-to-many connections will probably be the most common connections from a user viewpoint. This allows the user to have apparent access to a full range of resources via the wireless LAN technology. This will allow the coupling of the user to a wide variety of systems resources while still retaining the mobility and flexibility of the end user. The one-to-many connections will be defined by the user operations and actions and controlled by an administrative management process that is built into the overall network control level of the system. As one-to-many may use a combination of wireless and wired connections, the control levels will need to be logically rather than physically defined.

Many-to-many operations

Many-to-many operations would consist of several mobile user clients coupled to a variety of resources over a wireless LAN. The many-to-many operation is an extension of the one-to-many situation. It will again add complexity and management requirements to keep the users straight and their service requests properly addressed and handled. As multiple mobile users are being supported in most organizations, this form will end up as the most typical form of connectivity.

The major aspect of many-to-many couplings will be the management of the requests and returns of information to and from the mobile end users. The services may be performed via a mix of wireless and wired networks from resources that could be anywhere in the organization's networked world. Many-to-many management will have to keep track of all users and their network activities at all times and under all loads and conditions.

Wireless receiving devices

The wireless LANs will mostly support the layer between the end user and the higher orders of the information resources of an organization. In some cases traditional desktop personal computers can become wireless devices by adding a transceiver board instead of a wire connection. However, the more popular and heaviest uses of wireless LAN technology is likely to be in support of new user interface devices that are built for the wireless world. The following sections define some of these devices.

Laptop\Notebook computers

The evolution and popularity of laptop and notebook computers provide a ready platform for the interconnection to a wireless network. These devices are highly portable, battery run, and built to go anywhere, and run anytime in the hands of a mobile worker. They are being built with one or more standard Personal Computer Memory Card International Association (PCMCIA) facilities. These small credit card-sized devices can contain modems, storage resources, transceivers, etc. to support interconnectivity to a wireless network.

Laptop and notebook computers are also an accepted commodity in many jobs. They can support a first-order one-to-one connection and can be expanded to work in a one-to-many format. As the popularity of these devices grows and their production volume increases, their price will drop, increasing their popularity.

Personal digital assistants (PDAs)

The next level of miniaturization of computers will be the personal digital assistant (PDA). These units are handheld, calculator/cellular telephone-sized, battery-operated computers. Using touch screen or voice-activated technology to replace keyboards, the PDAs can put mobile workers into contact with a wide variety of information support activities that are stored within their memories. By adding wireless LAN connections to the PDAs, the mobile user can have access to the total world of information. The only restriction would be that the information has to be formatted and sized to fit within the capabilities of the PDA.

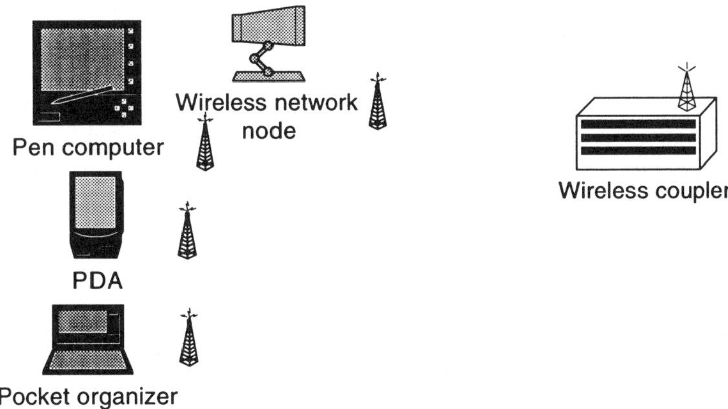

Figure 2.6: Various Wireless LAN Receiving Devices

PDAs are the nearly perfect receiving product for wireless LANs. They are highly portable but of limited self-contained capability. This makes them dependent on using a network to reach and interact with organizational information resources. Wired networks would certainly perform the required services, but they would remove the flexibility and portability of the PDA, restricting the key values the user is seeking. By using a wireless LAN connection, the user can increase the usability of the PDA as a window into the information resources appropriate to the individual user.

PDAs and wireless LANs will make a most appropriate marriage. Each needs the other to ensure their success. As PDAs move to more capabilities and power, they will include built-in wireless LAN connections which will ensure this bonding of technologies and services.

Wireless resource interfacing

In addition to personal user connections, the wireless LAN receivers can be attached to fixed service resources. One popular example is to have a remote printer that is shared via a wireless connection by all users in the coverage area. The device will provide its standard resource services, but the users will be able to select and access the unit via the wireless network.

Similar connections could be made to data recording stations and fixed information servers that contained local reference information. An example would be a wireless product or warehouse location server that provided query, access, and return information to a community of mobile users within a defined building envelope space. Whenever the user made a product request, their personal unit would send a fixed message to the wireless server, which would perform the required lookup and return the information to the mobile user unit.

Distance limitations

Local area networks are by definition limited in their coverage area. The use of a wireless medium will impose other limits and restrictions on the distances and geography of the network. The use of air transmission signals for the passage of information in the wireless LAN will be dependent on their strength, local interference, travel through the medium, contention and competition, collisions, and other characteristics of the physical and operational environment.

Typical distance limits for localized wireless networks are in the order of 80 to 300 feet between components. This can be extended with relay servers and other design alternatives. This limit is not unlike those of wired networks such as twisted pair (10BaseT), which is 300 feet between a unit and its wiring hub.

The distances for the open-space varieties of wireless LANs are the reverse of those for the localized forms. These systems can move information at higher speeds over distances that are tied to the physical line of sight through the air space. Some of these systems can move information over 2 to 3 miles of space, and others that use more powerful transmitters can send information over 15 miles. These forms of wireless LAN are far longer than the 1500 meter segments of a typical wired Ethernet and far exceed the individual link wiring of 300 feet for token ring networks. If one then moves into the arena of wireless satellite connectivity, the wireless LAN can couple an individual user in a remote location using a portable satellite dish to almost any point in the universe. There is no other technology that can cover this level of distance.

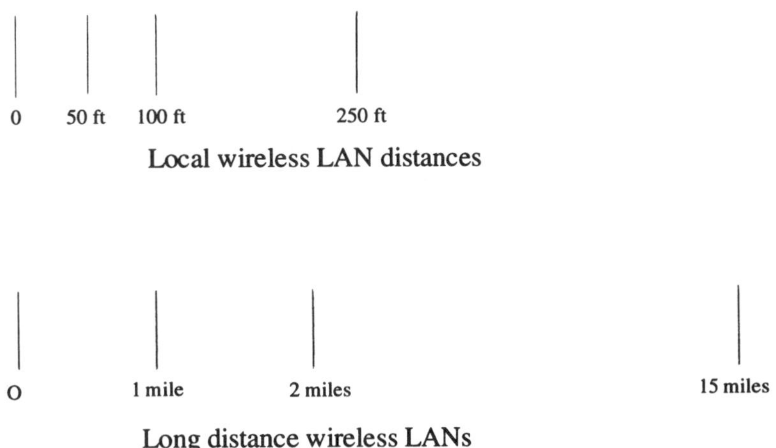

Figure 2.7: Wireless LAN Distance Comparisons

When viewed as a series of capabilities, the wireless LAN world is a large and diverse universe that supports limited to absolutely unlimited distances for connections and information transfers. Based on using the specific wireless technologies for an application, there may be limited distances for connectivity. However, by redefining the application and wireless technology users may find themselves in an unlimited distance wireless coupling. So the answer to limited distance connections is, it all depends on the application and technology chosen.

Distance limitations in wireless networks are more susceptible to the variations and uniqueness of the facilities around them. Walls, machinery, air contamination, competing wave motions, temperatures, etc. can all have effects (positive and negative) on the transmission of wireless information.

Speeds

Wireless LANs are not going to be as fast as those using wired media. This is because the air medium has more uncontrolled factors which can cause interference and reduce the reliability and accuracy of the transmissions. The trade-off is to run with lower speeds in order to be able to effectively combat the interference problems and deliver reliable, error-controlled services.

The nominal speed for wireless LANs will be in the range of 500,000 to 3,000,000 bits per second. The new 802.11 standard calls for a speed of 2,000,000 bits per second. Although 2 Mb/s is not the fastest speed for a LAN, several of the systems provide multiple channels of available frequency, allowing several users to have concurrent dedicated connections of a full 2 Mb/s bandwidth.

The new infrared transmission systems being developed by IBM are striving to establish new speed levels for wireless LANs in the 16 Mb/s range. This would make these systems compatible with the 16Mb/s token ring systems.

Although speed is important in any data communications and distribution system, the types of usage for wireless LANs may provide opportunities to constrain the loads and volumes of communications to fit within the limits of the technology and still offer good user services. This will be very application and usage dependent. It is accurate to indicate that wireless LANs will work better with limited amounts of data transfer, rather than as volume traffic systems.

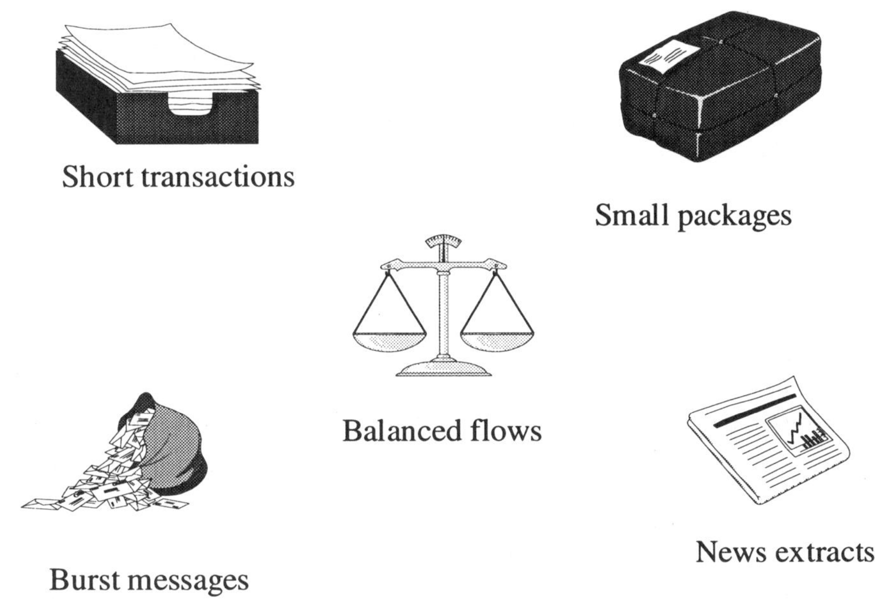

Figure 2.8: Limiting the Traffic Volume on Wireless LANs

Errors

Errors are the unpredictable factor in wireless LANs. Because users may be at different locations or mobile within their coverage space, the communications process is susceptible to interference and anomalies of the operating environment. The errors can be detected and retransmitted. However, too high an error rate will deteriorate the service and cause delays in service.

It is generally conceded that the error rate in wireless LANs is somewhere between two and five times higher that those experienced in wired LANs. The error rates are dependent on several factors other than random facility interference. They are dependent on the quality of the equipment, the installation, transceiver tuning, facility adaptation, user care, and a host of other elements.

Error-correcting protocols

Fortunately, technology offers several protective components to service the higher error possibility of wireless LANs. Among the most effective forms of error management are error-correcting protocols. These protocols have extra built-in data elements that are used both to detect errors in the data messages and to perform some level of automated correction of the message. If all else fails, the detected errors will be eliminated and a retransmission (with repeat validation) will be undertaken and repeated until the message is received with no detected errors.

Most of the popular error-correcting protocols are derived from the work with X.25 packet protocols. This is the basis for the new Cellular Digital Packet Data (CDPD) wireless protocol. Other approaches will use message cyclic redundancy codes (CRCs) that have been built for other transmission environments. Most of these codes are mathematically based, efficient to process, and effective in correcting the most common forms of data/message errors.

Security concerns

Anything that moves over the open air medium is susceptible to being picked up by unauthorized persons. If the information transmitted is mission critical or organizationally sensitive, then the wireless LAN adds an element of reduced security. There are three levels of response to the security issue. One is to avoid sending sensitive information over the wireless LAN. Another is to use an encryption-decryption process to convert the information to a form that is meaningful only to trusted users who have the decryption key. The third option is to use a technology that builds some security features into the system. This can be found in the multifrequency hopping of some systems, in that they move the communications to different frequency bands and only the transceiver on the network can interpret the order and content of the original transmission.

Security options for wireless LANs

There are several ways to approach the security issues in wireless LANs. These options are usually built into the protocols or are implemented via software specifications selected by the messaging process. Many of the operational packages that are provided by turnkey vendors for wireless LANs provide integrated security processing options such as data encryption. In such situations, the implementor would select these options and test their operation and then leave them as a transparent service for the user.

Like most components of a new technology, the security issues will be addressed by integrated service components that will be built in and transparent to the users. This does not mean the user should not be concerned; it only indicates that the protective layers will be included to maintain the level of capabilities available from competing technologies.

The security concerns for wireless LANs are real, but there are many practical answers that can make the overall systems no less secure than a wired system. As long as security is evaluated and planned at the design time, the wireless LAN can be considered an acceptable security risk for most situations and types of data.

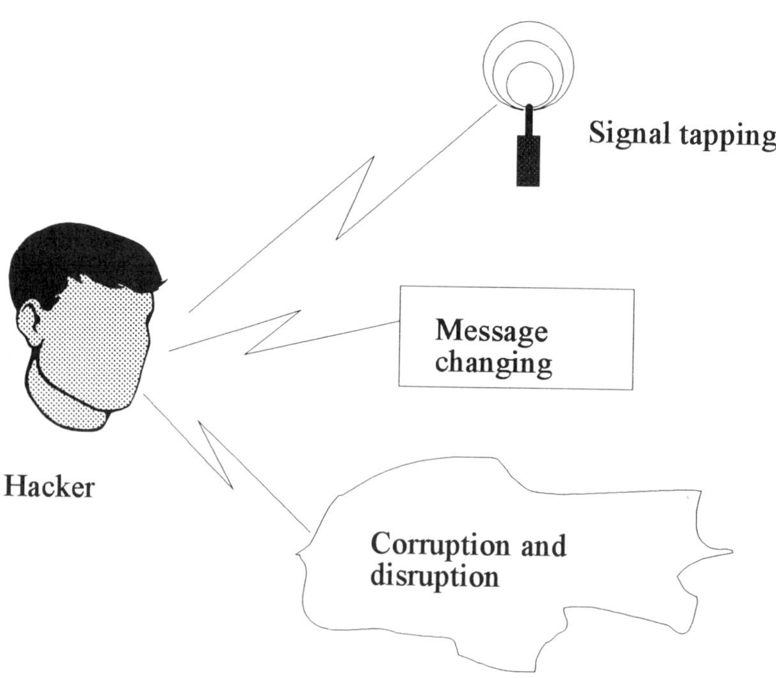

Figure 2.9: Wireless LAN Security Concerns

Interconnectivity

The key to wireless LANs is their ability to be integrated with other network systems and services. Wireless LANs can be used on their own in small and reasonably temporary situations. Their greatest use and success will be in building and supporting mobile user layers that can interconnect to other network layers and services. The wireless LAN interconnectivity will need to be reliable, secure, and transparent to the end users. The network to network interconnect processes should be accomplished in the hardware and software of the wireless transceivers. The interconnect should not require specific network instructions from the user nor should it delay or disrupt the flow of information over other network links.

CHAPTER 3

Wireless LAN Hardware and Software

Wireless LANs use various pieces of equipment that are unique to the network world. The transceivers, monitors, client units, and the interfaces between technologies are unique parts of the wireless LAN world. These technologies are new in their packaging, infants in their operation, unstable in their forms, and developing in their connectivity.

Wireless LANs also use the established components of networks and telecommunications to accomplish many parts of their operations. Hubs, bridges, network operating systems, servers, and other elements are often the same as those used in the wired LAN world. Some components may be extensions or modifications of wired products such as user interface software, backup servers, and transaction managers.

This chapter concentrates on reviewing the individual components of the wireless LAN world and defining their range of capabilities and services. Specific examples of products, technologies, and vendor offerings are presented as examples of the state of the wireless LAN world.

Equipment levels

The key types of equipment for wireless LAN systems are the portable/mobile client units and the transceivers that couple them to the wireless LAN environment. The portable units provide mobility at the user level. The transceivers provide wireless connectivity between portable and fixed level networks. Most of the other equipment is already a standard part of the wired LAN world. Because most wireless LANs couple to the wired LAN world at some level, the equipment levels are a full mix of the unique and standard network equipment.

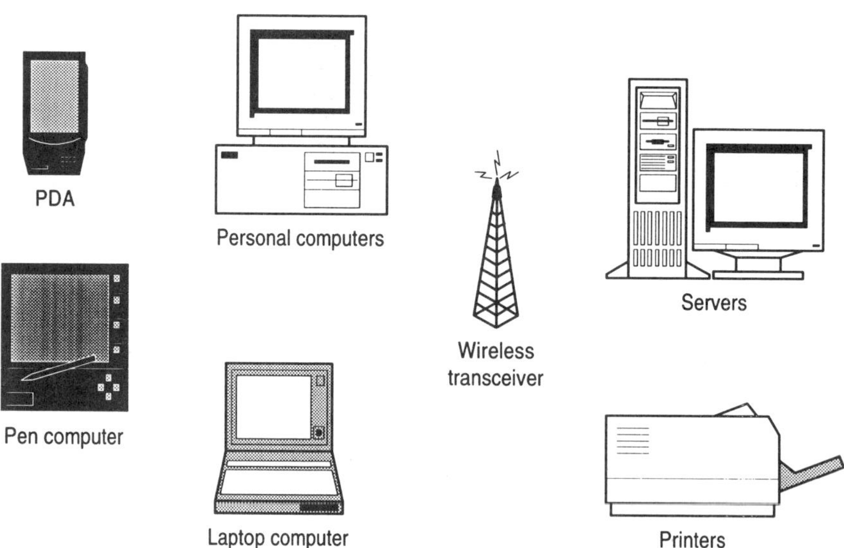

Figure 3.1: The Wireless LAN Equipment Spectrum

Wireless LAN equipment will need to build, package, transmit, receive, and interpret LAN messages and signals. The key element that makes a wireless communication part of the wireless LAN world is that it uses some of the standard LAN packet formats to service its transmission process.

Mobile equipment

The mobile equipment units for wireless LANs are portable computing devices that can act as full or partial client systems and interface to a wireless LAN communications unit. The typical range of units includes:

LAPTOP Computers—Standard personal computers in a laptop form using a wireless LAN transceiver unit usually attached to a standard PCMCIA slot.

Personal Digital Assistants (PDAs)—Small handheld (palmtop) computer units that support some type of communications capability.

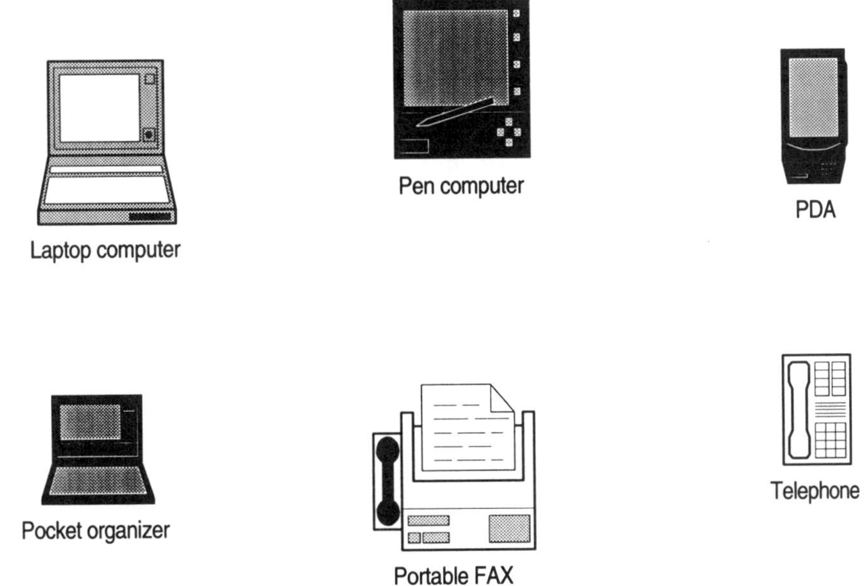

Figure 3.2: The Mobile Client Spectrum

Pen-Based Computers—Small handheld writing slates that can connect to a communications network

Speciality Units—Specially built (purpose-oriented) computing, data entry, and communications units that can perform specific functions and communicate to various network services. Included in this category would be radio frequency (RF) data collection units, portable bar code scanners, special data collection instruments, medical equipment, security control devices, etc.

Portable Bridges and Transceivers—Units that can support the interconnection from one LAN to another at the wireless layer. These can be PCMCIA transceivers, portable satellite ports, and other communication units that provide bridging services.

Server Interfaces

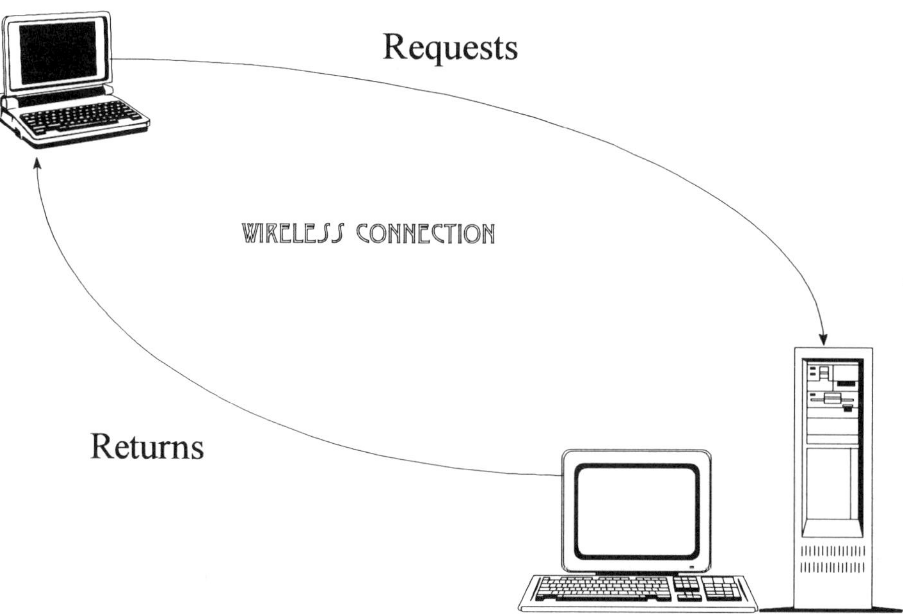

Figure 3.3: The Server Interfaces

The server interfaces will take the communications from the wireless LAN world and interface them to the fixed server world. The units will provide connectivity, transaction control, contention management, backup capabilities and other service and support functions. The wireless servers may be standard PCs, workstations, or minicomputers and even mainframes. The interfaces will take the requests from the wireless LAN users and present them to the server for processing or redirection to other servers.

Server interfaces will interconnect with the wireless LAN transceiver to capture the user transaction or request and deliver it to the server for processing. When the server returns the data to support the user request or transaction, it will hand the data to the transceiver, which will transmit the information to the proper user. These interfaces should be transparent to the user and provide an automated data traffic redirection service without user intervention. The server interface to the wireless LAN world may be a fixed interface address or may be a variable interface that covers various areas or space where the user is located when the user is in wireless contact with the server.

Peripherals and interfaces

The peripherals for the wireless LAN world are the same ones found at the client level on which the application is based. For portable/mobile computing units (client level) the typical peripherals include:

- PCMCIA cards
- Serial port couplings
- Parallel port connections
- Printers
- Communications connectors
- Keyboards
- Sound cards and speakers
- CD-ROMs
- Other portable-type peripherals

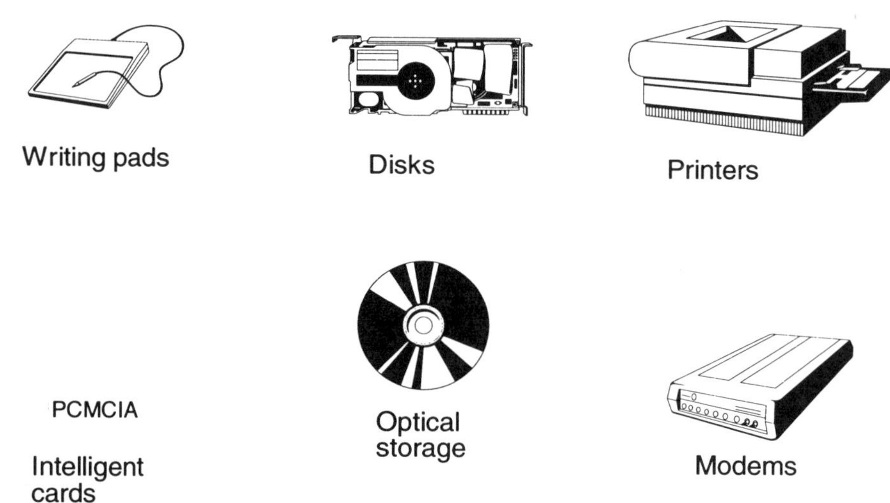

Figure 3.4: Wireless LAN Peripherals

The portable couplings for wireless units would include the full complement of portable user interfaces. These would include pen pads, touch screens, voice inputs, hand pads, and other new user interfacing devices. Most would be small, battery-operated units that support the mobility and personal interfacing of users to the data and information worlds.

The peripherals at the server level would be the same as those found on any server, with the exception of the wireless network interfaces. Mass storage, multiple user servicing, production output devices, high-speed dedicated input systems, and other service units would be found at the servers within the wireless LAN world.

Transceivers

Transceivers perform the transmit and receive functions within a wireless LAN environment. The units can take a signal request from a unit and transmit the data packet into the wireless medium. The receive portion of the unit can receive the transmissions, determine who they are for, and take those addressed to a specific station into its operating environment.

As dual-function devices, transceivers can operate in only one mode at a time (half-duplex). They may be tied up transmitting and will have to wait until they can perform a receive process. As the active process will in turn tie up the network connection, the half-duplex nature of the transceiver will not be a major concern.

Media components and conditioning

The media components of wireless LAN systems are the air through which the transmission waves will travel. Unfortunately, the air medium is subject to many variations and pollutants which cause deterioration in the performance of moving signals through the medium. In many settings the medium will require cleaning and conditioning to support a reliable signal transfer.

There is very little that wireless LANs can do to change the air medium through which their signals will travel. They must, however, constantly monitor the air quality and the accuracy and strength of the wireless LAN signals that are passing through the air. The continuous monitoring process will be build into the transceivers for the wireless connection. When the performance of the air medium fails or drops, the transceivers have to switch to alternate forms of operation to sustain the accuracy of the communications.

Some of the conditioning and signal control steps include:

•   Suspend operations

•   Switch to other frequencies

- Move to other forms of signaling

- Retransmission of the messages

- Multicasting

The key process is to get the messages accurately through the air medium to and from the client/server components. When the air medium is optimal, such services may be dormant; however, when the air quality deteriorates, these conditioning alternatives need to come on line automatically and expeditiously.

Wireless-to-wire connections

The wireless LAN will seldom be a self-contained environment in which all elements are coupled only via a wireless system. The normal situation will have wireless clients coupled to a server level, which is in turn coupled to the wired world. This allows the wireless user the best of both worlds. They will be able to be mobile and wireless from their client systems and still receive the benefits of connectivity to the wired world.

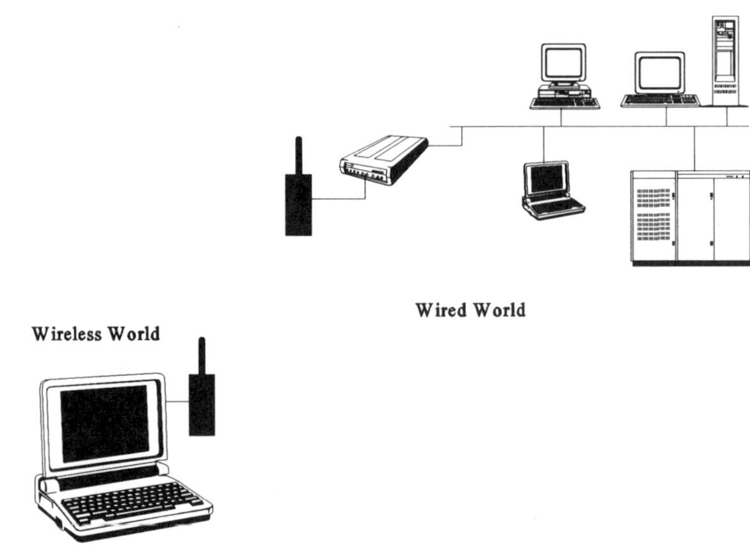

Figure 3.5: Wireless-to-Wired Couplings

Wireless-to-wired connections will involve signal recognition and translation from one form of medium to another. They should also be transparent and provide the users with information services that do not identify themselves as being handled by a wired or a wireless world. Full operational transparency where the wireless and wired worlds interact with seamless and transparent interfacing is a key measurement criterion for the wireless-to-wired coupling.

LAN to LAN Connections

Some wireless LANs will be stand-alone operations; however, most of them will involve coupling one LAN to another. The couplings may be wireless to wireless or wireless to wired. There may be many levels of LAN-to-LAN couplings as the wireless units become accepted parts of the enterprise network integration. In some of the couplings, the wireless LAN connection will become a key part of bonding other network layers together.

Although many wireless LANs will provide user-level services, others will be devoted to the interconnection of one LAN to another. For example, wireless LAN bridges can supply long-distance, high-speed, wireless data connectivity between one LAN and another. At the lower speed end of this spectrum (between 1 and 2 Mb/s) these wireless LAN bridges will compete with T1 communications lines but with less cost and simpler LAN-based administration. At higher speed levels (16 Mb/s and above) the wireless LAN couplers will compete with T3 lines and other services such as frame relay.

Key factors in the use of wireless LAN couplers will be support for longer distances (up to 15 miles) with lower costs (no telephone company line charges) and the use of internal LAN management techniques. Wireless LAN-to-LAN couplings may become a major playing card in the building of cost-effective enterprise networks.

Software layers

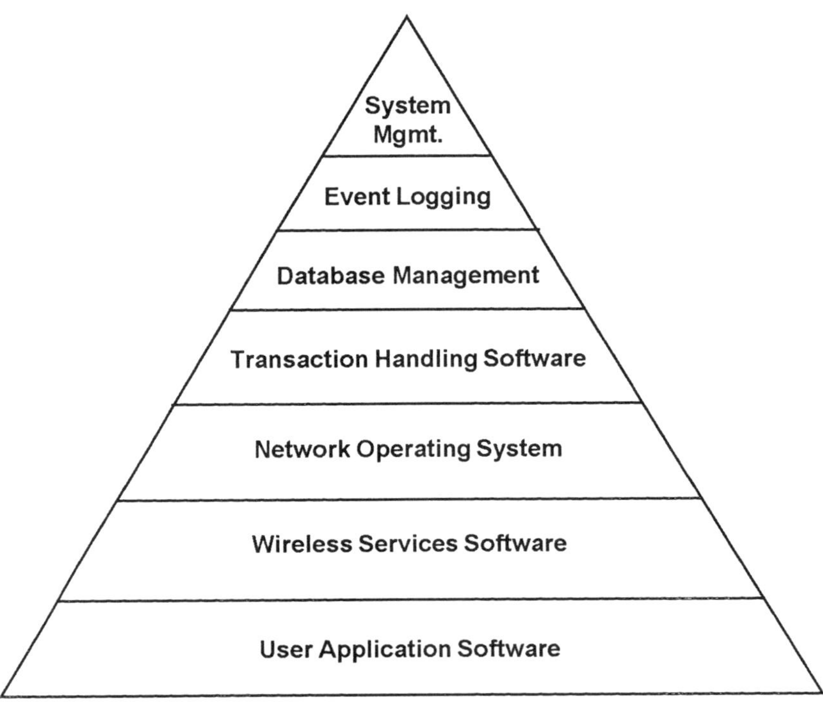

Figure 3.6: The Wireless LAN Software Layers

The software layers of a wireless LAN system are similar to those of a standard wired LAN. The software components include:

- Network operating system (NOS)

- Client operating system

- Server operating system

- Wireless utilities software

- Application software

- Utilities

- Systems management software

- Security control

The software layers in the wireless LAN world can be supplied by products that are proven in the wired LAN world. The main additions will be at the user interface level, where the portable/mobile devices will be new to the systems equipment complement and may use special operating and applications software.

The major software functions in support of the wireless LAN will be the formulation of the network messages, the interfacing to the wireless network, the access links to the database, the display of user input forms, the presentation of the network-delivered data elements, and the status and error messages on incomplete operations.

Bridges, routers, and gateways

The interconnection of wireless, wired, and other network environments will utilize internetworking components such as bridges, routers, and gateways. These components either will operate at the wired network level or will need to be modified to function in the wireless world.

Bridges will provide the interconnection from one network to another. A wireless bridge would interconnect two separate network environments, which could be wireless or wired. The routers will perform the analysis of data packets and route them to the proper network environment. Gateways will exchange the network protocol package from one environment such as an open system interface (OSI) LAN to another such as an IBM systems network architecture (SNA) LAN.

The purpose of the bridges, routers, and gateways is to provide interconnection services so that short-length, local area networks can be integrated and extended across the enterprise. The units that support wireless-to-wireless interconnectivity will have to conform to the signal formats and waveforms of the wireless LANs.

The fact that wireless LANs are introduced into the internetworking architecture should not change the functional components or layers of the LAN-to-LAN equation. The wireless couplers will adhere to the standard LAN protocol formats and will differ only in their usage of the physical (OSI level 1) medium. A wireless Ethernet unit should be able to couple to a wired Ethernet system at some point where the media form of the signals is transferred from wireless to wired. The data packets and the overall control mechanism will still conform to the full International Electrical and Electronic Engineers (IEEE) 802 series standard for the specified type of LAN.

Equipment selection criteria

The selection of wireless LAN equipment can be supported by using the following evaluation criteria:

- Functionality
- Flexibility
- Error management
- Scalability
- Expandability
- Performance
- Vendor support
- Software facilities
- Integration support
- Interoperability
- Ease of setup
- Ease of use
- Application interfacing
- Management services

The evaluation of the above criteria can be done with a weighing value and a scoring matrix. Each option should be compared based on the major criteria and then evaluated against other products.

CHAPTER 4

Radio Wave LANs

The use of the radio waves for the wireless LAN medium is the simplest and most logical way to interchange information between local computers. The problem is that all radio frequencies are carefully managed and controlled by the federal governments. In addition, most of the radio frequency spectrum is already in use by other applications and assigned clients. Radio wave LANs have been granted usage of three frequency bands that are small unused niches in the spectrum. Known as the spread-spectrum areas, these frequencies can be used without license for low-power, limited-distance communications that stay within the confines of a building.

Radio wave LANs provide a broad range of flexibility and portability. They require little tuning and can be used within a reasonable area. They are susceptible to local interference from equipment and other electrical devices.

Bandwidths

Radio wave wireless LANs operate in the 900 MHz range. They utilize from one to three segments of this band to transmit and receive signals between units. The bandwidths are very limited and are in the ISM band. This band is known for its susceptibility to interference from radar, microwave ovens, medical equipment, and other devices.

The 900 MHz band space is most notably used by today's portable telephones. These are the in-home units that are portable within a limited space and transceive their communications to/from a base unit that is coupled to the telephone wire system. The frequency is different, so there is little conflict in the transmissions, but the similarity in the transmissions and the typical problems found in the portable telephone systems are carried over to the radio wave wireless LANs.

<table>
<tr><td>902 MHz</td><td></td><td>2.4 GHz</td><td></td><td>5.725 GHz</td></tr>
<tr><td>To</td><td></td><td>To</td><td></td><td>To</td></tr>
<tr><td>928 MHz</td><td></td><td>2.484 GHz</td><td></td><td>5.850 GHz</td></tr>
</table>

100 milliwatt (MW) --- 250 MW --- Limit 1000 MW (1 watt)

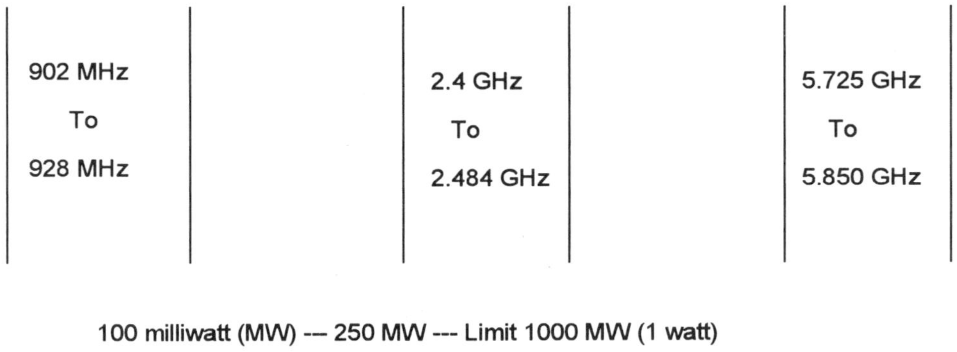

Figure 4.1: The Spread-Spectrum Bandwidth

License requirements

The good news is that no license is required in the spread-spectrum bands as long as power is kept within limits and the signal stays very local. The bad news is that the frequency space is limited and the power must be kept very low.

This may change as the Federal Communications Commission has been considering requiring licenses for the spread-spectrum bandwidth. They have had little luck in the past imposing a license requirement on previously unlicensed elements; this case is likely to end up the same.

Spread-spectrum

Spread-spectrum bandwidth is in the 902 MHz to 928 MHz range. Additional bands are in the 2.4 Ghz to 2.484 Ghz and the 5.725 Ghz to 5.85 Ghz ranges. Although limited, these bands are covering only a limited distance and can carry considerable traffic if managed properly. Some of the spread-spectrum space is the same as that used by the newer portable telephones found in many homes. They use the 900 MHz channels and can transmit their voice calls up to 200 feet from the line-based transceiver.

Single versus multiple channels

Radio spectrum wireless LANs can split the frequency up into multiple channels, or they can use a single channel and dedicate the total bandwidth. The trade-off involves load transfer speeds versus the support for simultaneous intercommunications. In a multiuser situation the choice would tend toward multiple channels to increase the availability of services. In small groups, the choice might be to favor a single channel to allow for faster movement of heavy traffic.

The choice of channels will vary with the situation and the users. The choice can be changed at the transceiver setup levels. However, the changeover takes some time and does not favor an instantaneous load condition change. Usually the specification of approach will be made at the time the system is planned and configured. Many of the transceivers can operate over multiple channels, allowing the decision to be deferred until the installation. It is also possible to try several frequencies within the building space to determine whether one performs better than others. It is also possible for the LAN administrator to make the switch to a new frequency at a later time, if the previously selected one becomes troublesome.

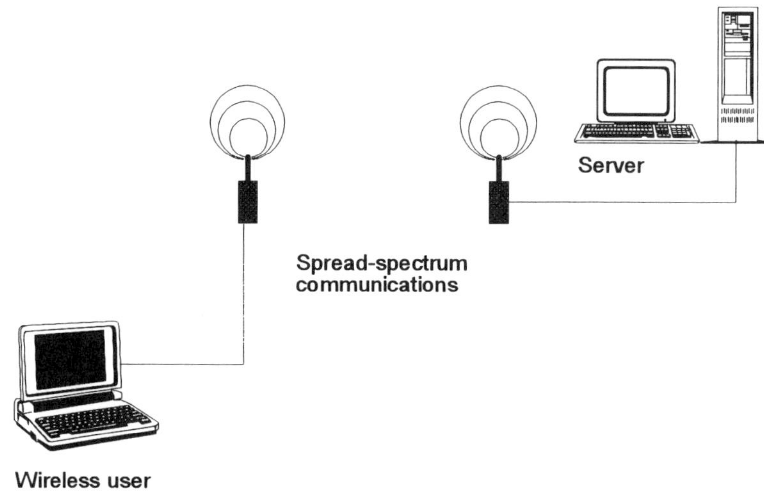

Figure 4.2: Spread-Spectrum Operations

Frequency hopping

Communications on one channel of the radio spectrum can be transferred to another available channel as the traffic passes from one radio zone to another on its way to the final destination. This is called frequency hopping. The frequency hopping is performed automatically as the signal is received and retransmitted across the network nodes. The process is very similar to the one used in the cellular network as a user passes from one cell into another.

Frequency hopping has two basic advantages in wireless LANs. One is that the system can use the available bandwidth more efficiently. The second advantage is that frequency hopping makes it very difficult to eavesdrop on spread-spectrum data traffic, because only the transceivers know which channel is carrying the next block of communications. By the time an eavesdropper was able to shift to another channel, the communications traffic would be gone.

Frequency hopping is built into the transceivers of a spread-spectrum wireless LAN. It would be selected as an operating option at the time of installation and setup. Many of the systems operate in this mode as their default state.

Direct sequence coding

The passage of information through the wireless LAN network will move across the network nodes using the spread-spectrum bandwidth. Direct sequence coding happens when the traffic is directed between specific nodes for passage into other network media. For example, the user may need to travel over a wired network to reach a remote server. The direct sequence coding would send the wireless communication to the nearest wired LAN station and then move the request into the proper format for the wired system.

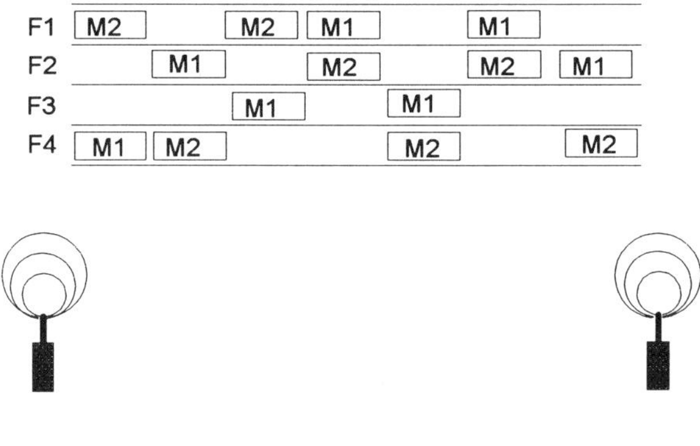

Figure 4.3: Frequency Hopping versus Direct Sequence Traffic

Directional and omnidirectional antennas

The connection of a mobile client to a wireless LAN will require a physical device to generate the waveform pattern that will travel through the air/space medium. Various types of antennas will be used to fulfill this physical medium conversion step. Some of the antennas will be directional. These will require tuning the wave beam from the transmitter to the receiver. The tuning may require space entry (beaming microwaves up to the ceiling level) or it may require direct line of sight (infrared couplings).

Omnidirectional antennas will simply beam the wave transmissions up into the space around the transmitter and the transmissions will radiate uniformly in all directions around the antenna. The pickup receiving antenna need only be within the wave transmission area of the signal to receive and convert it.

Longer distance bridging

Wireless LANs are mostly used for limited distance, local services. However, it is possible to use the same concept with higher power broadcasting levels and travel over several miles. The Air-Bridge system from Persoft has the capability to extend LAN-based devices and services to distances of over two miles. The Air-Bridge maintains 2 Mb/s speeds between the remote connections. The LAN-LAN wireless connections offer higher speeds than wired telecommunications facilities at lower costs. In addition, they can be managed as part of the internal network process without having to interface to external service vendors. Other examples of these products include Air-Link Radio Modem, Solectek Wireless Bridging, Silcom Freespace, Black Box LAN2 Bridge, and many others.

Transmit facilities

The transmit facilities for wireless LANs involve creating waveforms within the atmosphere that are capable of moving through the "air" medium to couple one location with others. The transmitters must take the data signals and build logical transmission packages (packets) and then generate the appropriate waveforms in their local space.

The transmitters for wireless LANs must be small and portable. They must operate from batteries and have a long enough life to remain operational for periods when they are not in contact with recharge energy. The transmitters must also be relatively immune to problems of alignment and tuning. They will be operating in a variety of mobile situations and must remain reliable in terms of both positions used and the continuity of message throughput.

Receivers

Receivers perform the reverse of the role of transmitters, in that they receive the signal waveform from the atmosphere and convert it into usable data for computer processing. The receiver must operate on the same frequency as the transmitter. The two units are usually integrated into the same computer circuit card and bundled as a transceiver (transmitter-receiver). Unfortunately, the standards for wireless transceivers are not yet defined, so units from one manufacturer will work only with like units from the same firm. Many units are even model and revision level locked into working with only like units of the same model and revision level. This may change over time, but it is not unlike the early days of many other technologies such as Ethernet and facsimile.

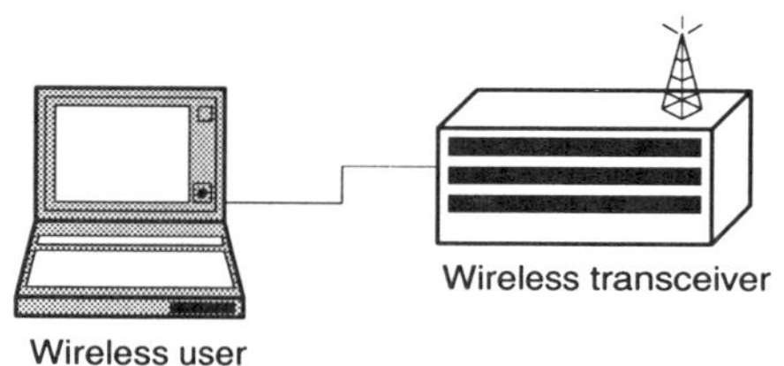

Figure 4.4: Wireless LAN Receivers

Receivers have the more difficult job in the work flow between transmit and receive. The transmitter simply places it signal package into the medium and rests until called upon to do it again. The receiver must receive the signal, validate the address, and then check the validity of the content. If errors are discovered, the receiver will be responsible for initiating the retransmission and the successful completion of the acceptance of a good message.

Receivers must deal with varying signals, dropouts, distortion, conflicts, fades, etc. The sensitivity and built-in hardware logic applied to the processing and authentication of the correctness and accuracy of the received signal are part and parcel of the receiver's capabilities. In addition, the direction of the received messages and conversion into usable computer data are part of the receiving responsibility.

CSMA/CD processing

The passing of messages in a wireless LAN is usually done with the Ethernet protocol. This protocol accesses the wavelength channel and tests if it is busy (carrier sense, multiple access); if it is not busy the unit begins transmitting. If the channel is busy, the unit will pull back and wait and then resample again to see if it is available. If two units acquire the channel when it is apparently not in use and begin transmitting, their signals will collide, causing errors. The protocol provides collision detection as a mechanism to pull back and wait random periods before resuming access and transmission on the channel.

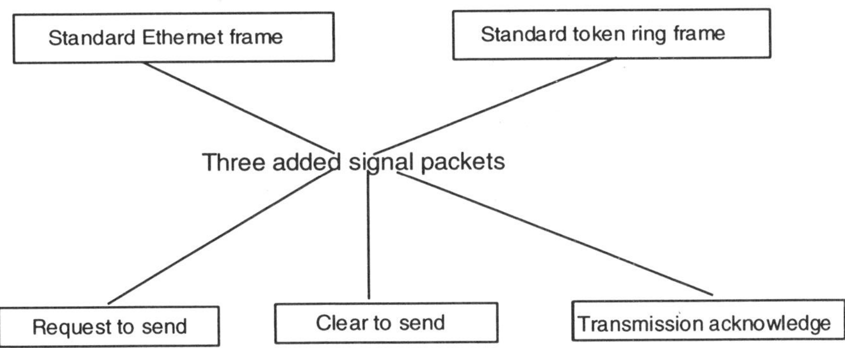

Figure 4.5: The Wireless Ethernet Frame

The Ethernet (CSMA/CD) protocol is simple and passive in form. This makes it easy to support in a wireless LAN environment. The primary limitation with CSMA/CD is that it cannot sustain heavy loads. However, most wireless systems are limited in the number of users, which restricts the loads on the network. The alternatives for heavy loads are to use dedicated (nonstandard) protocols or split the CSMA/CD network into more segments to configure it to suit the traffic levels.

LAN interconnections

The interconnection of a client or server unit to a wireless LAN is made by connecting the computer to a wireless LAN interface transceiver card. This coupling can be via an internal connection to the computer bus (usually using a card slot). It can also be connected via an external port such as a parallel or serial port that is a standard interface on most computers.In many situations the client computer is coupled to another client computer in a series of local loops. This can be via a daisy-chain or a small local area wired network. In a small work group this would provide a way to reduce the number of wireless connections while keeping the wired costs low. When users wanted to communicate beyond the local loop they would request access to the wireless link which they shared with the other local users. This would have the effect of increasing the utilization of the wireless link (four users rather than one) while keeping the wireless connections less costly per user and reducing the interference of multiple wireless communications occurring in the same air space.

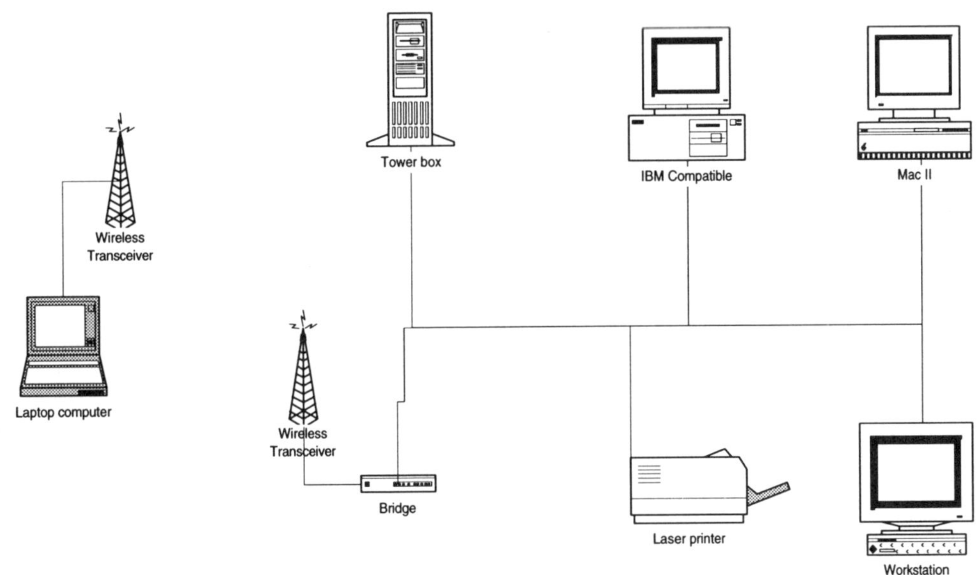

Figure 4.6: Wireless LAN-LAN Interfacing

Another approach is to use a wireless bridge as a dedicated service that will transfer signals from one network to another over the radio link medium. The LAN traffic would move to the bridge, which would convert the traffic to radio frequency with enough power to cover a range of 3 to 30 miles. The receiver would be tuned to receive the signals at the remote location and convert the radio waves back into internal LAN traffic.

Bus and channel cards

The computer bus or microchannel (in the case of IBM PS/2s) is a high-speed direct connection to the other processor cards of the computer unit. The bus couplings often have from 50 to 100 signal paths and represent the highest available data transfer speeds outside the motherboard and the main processor chip. There are several different buses with different configurations and speeds. These are usually related to the main CPU chip and the architecture of the particular computer unit. The popular bus choices include:

- ISA—Instrument Society of America—an older 8 and 16 bit bus with data transfer speeds in the 6 Mb/s range.

- EISA—Extended ISA—a newer 16 and 32 bit bus than can attain data transfer speeds of up to 32 Mb/s

- PCI—Peripheral Component Interchange bus—associated with the Intel Pentium chip. A 64 bit bus that can attain maximum theoretical speeds of 132 Mb/s.

Parallel and/or serial port connections

Most computers can also be attached to wireless LANs via their existing external ports. The preferred coupling is via the parallel port because it is faster and has more control lines than the serial port. The parallel port is used extensively when a portable laptop or notebook computer is coupled to a wireless LAN. These computers have no usable bus slots and therefore cannot take a standard network interface card.

Fortunately, the parallel ports on personal computers are highly standardized, making it relatively easy to produce connector boxes that can feed from the parallel port to the wireless LAN transceiver. There are two problems with both the parallel and serial port usage for a LAN. One is that there is no power available on the port lines, necessitating a separate power supply or connection to another power source (such as the keyboard port). The second one is more endemic, in that the speeds of the serial and parallel ports (although variable) are limited. The serial port stops at about 115 kb/s and the parallel port can be driven at about 600 kb/s. Both of these are slower than most wireless LANs.

Hubs and access points

The building of most LANs involves more than coupling the computer together. The physiology of the system is usually supported by a variety of connecting hubs where various local units are wired to a common point and this point is connected to other hubs for information transfer and management. These hubs may be active or passive and form access points between the user and the network service layers.

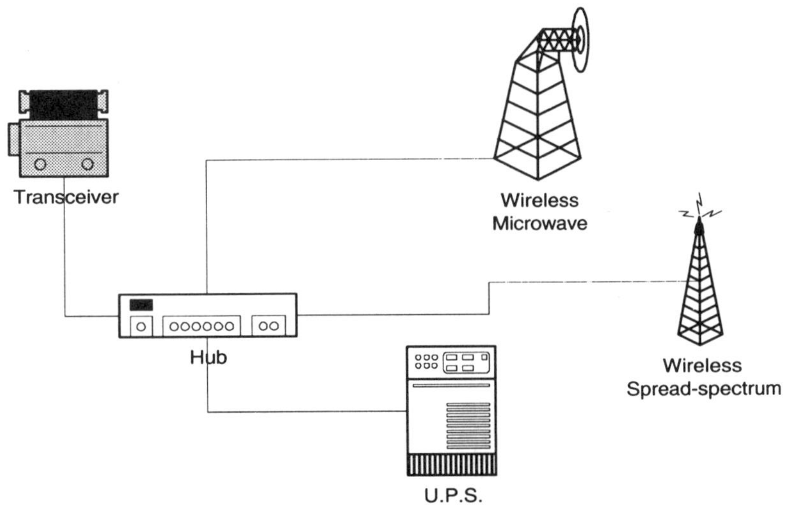

Figure 4.7: Wireless Hub Couplings

Hubs are devices which are used to couple various units of a network together. They represent the center hub of a wheel where the spokes couple together the outer part of the wheel to the shaft. In a network the hub is the place where signals can be brought (from the outer spoke locations) and redistributed to other spokes in the network. One hub can be coupled directly to other hubs (like the shaft in the wheel, which goes from one hub to another).

Hubs often make convenient locations for the connection of servers and/or other shared resources. In a wireless LAN a hub would be a logical place to make the bonding from the wireless LAN to a wired LAN. The hub would also be the best place to put sharable resources and servers that are needed by the wireless unit clients.

Software connections

All LANs require some form of software connection into the computer units on which they are operating. LANs are still an unnatural extension of a basic computer system and must be integrated with the systems using software programs.

The use of a network operating system (NOS) is the usual choice for bonding the LAN services into the normal computer operations. The NOS will talk to the native computer operating system (such as DOS, Windows, OS/2, Apple System 7.5, or UNIX), and extract the requests for LAN services, and oversee their execution and delivery.

The NOS has provisions for establishing the identity of the LAN services and then performing interrupts on the native operating system to transfer requests for these services from the computer to the LAN interface.

The NOS is like an overlay manager that watches the native operations in the computer and entraps any request for the LAN, while allowing standard internal requests to pass by to the computer operating system. This process is known as redirection, and the routine that performs this act is known as the redirector.

The entry to the NOS requires that the applications software be equipped to be able to ask for services from the LAN world. Most stand-alone software asks for services only from the single computer on which it operates. LAN-compliant applications software can ask for local and LAN remote services. LAN applications are often called multiuser software because multiple users can access the resources of the network. Again, the LAN services are provided to the applications software via the NOS, so a LAN-based system requires both the NOS and some part of the applications software to be running on all connected units. The application software must be able to interoperate with the NOS to obtain network services.

The multiuser and NOS software layers add costs, resource requirements, and processing time to the systems equation. They also often become bottlenecks as multiple users or application activities request services from the LAN and the resources available through it.

Waveforms and encoding

The interaction of a computer unit and the wireless LAN requires the requests for services and the data exchanges to be encoded into a waveform that can be handled in the wireless environment. The waveforms for a wireless LAN are energy patterns at selected frequencies that can be passed through the physical space medium (air, physical objects, bodies, etc.) These waveforms must be able to hold their shape and content and travel their specified distances in a reliable fashion. The information to be transferred is encoded into the waveforms so that it can be extracted and translated into computer-processable requests at the receiving end.

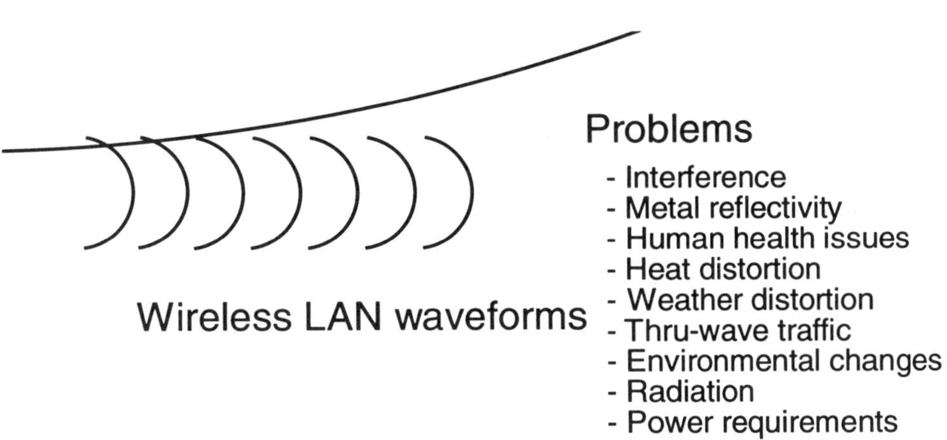

Figure 4.8: Wireless LAN Waveforms

Data rates and throughput

The radio wave wireless LANs can sustain only limited data rates and short bursts of throughput. These limits are due to the error-prone nature of the wave media. If the waves can be made more reliable (use of higher power) and of shorter form (more information transferred), then these limits could be removed.

The data rate and throughput capacities of wireless radio LANs are in the range of 600,000 to 2,000,000 bits per second. These could be raised to the range of 4 to 6 Mb/s if the signal forms can be enhanced.

Errors

The errors in radio wave wireless LANs are usually associated with interference that modifies the data streams while they are in the transmission medium. The errors can cause changes or dropping of information bits and thus invalidate the received packets.

Error management and control is needed on all radio wireless LANs. It can include the acknowledgment of all transmission packets or the use of a sampled evaluation technique.

The errors in radio wireless LANs are directly related to the environment in which the system is operated. The more transient noise that is in the environment (such as a factory or industrial setting), the greater the error rate. In some settings, such as an office, the error rates are more variable. Often the close proximity of other computers, copiers, telephone systems, video systems, fluorescent lighting, etc. can cause more interference in the radio signals than heavy industrial machinery.

Error correction

Error correction techniques in radio wireless LANs usually consist of packet error checking and correction. Retransmission is done for all error packets that cannot be corrected using the error-correcting codes. This is embedded into the wireless LAN software and the user does not need to worry about it.

Encryption

Because the radio wave signals are sent over public space, there is always the possibility of someone capturing the signal and copying it for improper use. One approach is to avoid sending any critical, confidential, or restricted information over the wireless LANs. Another is to apply some form of encryption algorithm to the data portion of the signal packet and then have the proper receivers possess the decryption key so they can extract the true contents of the data from the encrypted packet.

Encryption in wireless LANs would be no different from applying encryption in other network layers. The encryption process is either available as part of the NOS or will be linked to the NOS as a separate off-the-shelf software package. The selection and setup of the encryption and decryption mechanics would be done as part of the design and implementation of the wireless LAN system.

Radio wave competition

Radio wave wireless LANs will be using a small amount of allocated bandwidth in the radio spectrum. Other radio services such as portable telephones, pagers, and similar devices will also be using nearby bandwidths. This could cause some interference with or competition for the limited available space. Care will have to be taken in the planning of a radio wave wireless LAN to be sure that there is minimum contention for the bandwidth within the space to be supported by the wireless LAN.

CHAPTER 5

Infrared LANs

Infrared LANs use the nonvisible portion of the light spectrum. The wavelength is in the range of 900 nanometers. This is the spectrum used by remote controllers, TV remotes, and some security detection units. The wavelength provides short signals with good resolution but very short range, with tuning (sight alignment) required.

In the LAN operation the infrared systems will use a broader spectrum of wavelength to provide a less critical tuning connection between the LAN nodes and the network transceivers.

Light beams

The infrared beams are waveforms just outside the visible light frequencies. The beams are still generated frequency levels in this spectrum, with their data contained in pulsed encodings. The sending and receiving are different from those at radios frequencies because the light beams are unidirectional rather than omni-directional as they leave and enter sending and receiving units.

Sending

The sending of infrared signals requires the network operating system (NOS) to interrupt the request for LAN services and send the data to the network interface unit. The interface unit will package the data into LAN packets and engage the infrared transmitter. The transmission is aimed at the receiving unit somewhere within the tunable space. The infrared signal is received and either decoded or passed through a retransmission to another (addressed) unit in the wireless LAN configuration.

The sending of infrared signals can be a direct one-hop transmission or it can be set to be multihopped between transceivers until it reaches the specified server or user in the wireless environment.

Receiving

The receiving of infrared LAN signals requires that the receiving sensor be located within the line of sight of the transmitter. The angle can be fairly wide ($\pm$ 30 degrees), but the beam must reach the sensor with adequate strength and continuity. The straighter the beam and the more directly it is aimed at the receiving sensor, the better the reliability and throughput of the information transfers.

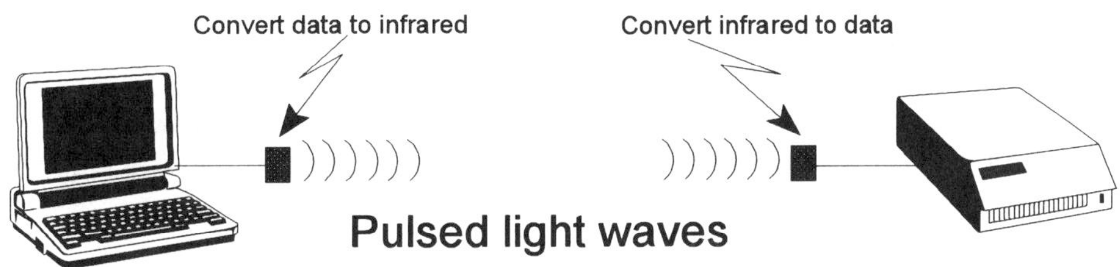

Figure 5.1: Infrared Light Beam Transmissions

Optical frequencies

The optical frequencies used by the infrared LAN signals are just below the visible lightwave spectrum. These light beams are not visible to the naked eye and they are not harmful to humans. As part of the optical spectrum, these signals are immune to electrical interference and they are relatively immune to eavesdropping because of their tuned send and receive nature. The light signals can also support a relatively high speed of data traffic within their bandwidth.

Speeds

The infrared signals can be operated at relatively high speeds. Current laptop-to-laptop infrared services are running at 16 Mb/s. The theoretical limits of the technology are up around the 100 Mb/s range. This can make the infrared services the fastest of the wireless LAN technologies.

Line-of-sight beams

The infrared LAN signals are directional signals. The radiation pattern is focused from the transmitting source in a straight line form in the direction in which the beam is pointed. The signal pattern spreads out (disperses) as the signal travels; however, the spread at the end of the receiving distance is about 3 to 4 feet. This means that the beam must be in line-of-sight alignment from the transmitting source to the receiving point.

The receiver for the infrared beam can be mounted in any location where there is a clear path from the transmitter to the receiver. A common practice is to mount the receiver in the ceiling in a corner of the usage area. This makes a convenient pointing space for the user to aim at with a high likelihood of reaching the target without interference. The diagram in Figure 5.2 depicts a typical infrared beam space layout.

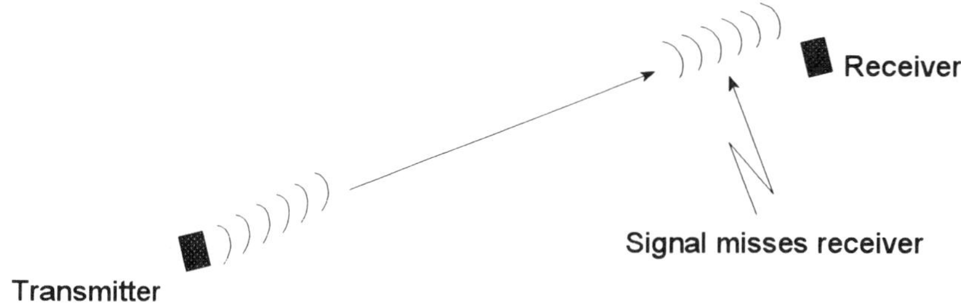

Figure 5.2: Infrared Directional Beams

Power requirements

Infrared power requirements are very low. The infrared remote controllers use small batteries that last for several years. This makes infrared ideal for portable, battery-operated computer units.

Interference

Infrared signals are very susceptible to physical forms of interference. If a human body enters the infrared beam space, it will absorb the signals and the receiving antenna will be blocked. Other forms of interference can be caused by signal deflection, bouncing signal interference, and a host of other physical space distortions.

In most successful infrared installations it is important to get the infrared signal up into the higher facility spaces where there is less likelihood of physical interference. Even this move can be subverted unintentionally by strange occurrences. One large open office environment had a very successful infrared LAN installation, except the signals would distort and fail around the user environment once each day. The cause was finally traced to a mail delivery shopping-type cart that had a high pole attached to it with a metal flag which was used to look over the partitions and spot where the mail cart was in the environment. Once the metal pole and flag were replaced with wood and cloth, the interference disappeared.

Errors

Errors in the infrared transmission of LAN data are most likely to occur in the coupling of the transmitter and the receiver or in local interference within the line of sight of the transmission. The receiving unit will review all packets and check their structural validity. Any packets found to be in error will be requested to be retransmitted.

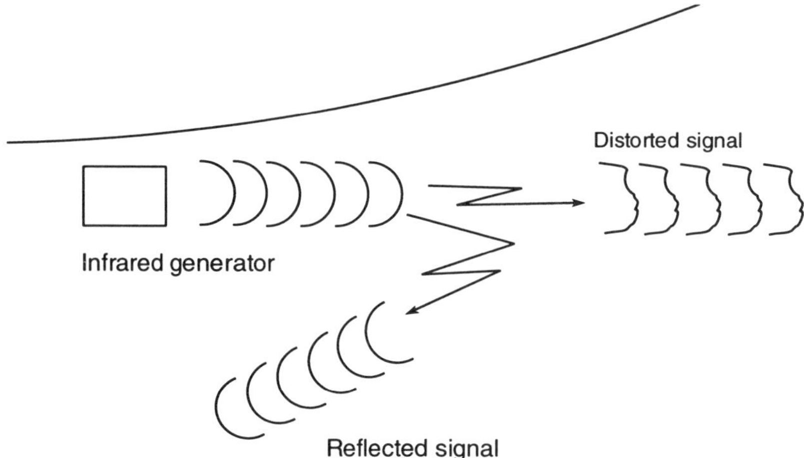

Figure 5.3: Infrared Errors

Short-haul infrared

Infrared transfer services are also being used for very short distance intercommunication. Many new generation laptop computers are coming with built-in infrared transmit and receive capabilities. IBM, Texas Instruments, Hewlett-Packard, and others are making infrared the first level of interconnectivity for their laptop computers.

When a laptop user needs to move data to or from the laptop computer to another computer (such as a desktop unit), the user would simply position the devices within 3 to 8 feet of one another, perform some send-receive tests, and then open up the systems for rapid transfer of the chosen data and information. The data movement speeds for this form of infrared wireless LAN are in the region of from 600 kb/s to 16 Mb/s. The key is to support the bus speed of the slowest unit (usually the laptop) so that it forms an almost continuous service capability between its internal resources and those available via the infrared LAN.

The short-haul infrared wireless LAN operates very well in the eyes of the user. The network is one on one with little or no contention from other users. The resources of the desktop system are also dedicated to the support of the client unit, while the linkage is in operation. Thus the infrared wireless LAN greatly increases the power and resources available to the portable client-level system and provides exceptional performance at the same time. In addition, the laptop user can access the enterprise resources through the LAN connection of the desktop. This also keeps the LAN management link under the control of the network manager, who does not have to worry about portable users, because their port of entry through the enterprise LAN is through the desktop. This in essence makes the desktop a firewall for the rest of the enterprise world.

The future of infrared

Infrared communication has several positive characteristics. It is low power, requires no license, needs an area tuning, but provides some security from eavesdropping. The short distance is not a problem when the units are in close proximity anyway. Infrared is the least expensive and most reliable form of short-haul wireless interconnection. It will continue to grow to become the connection of choice for personal digital assistants (PDAs), laptop computers, and other portable data collection and service units.

CHAPTER 6

Microwave LANs

Microwave wireless LANs use the frequencies in the 18 GHz to 24 GHz range to encapsulate and transfer the LAN data packets. These frequencies are in shortwave form and are very high. They will pass through most things without loss or distortion. This makes microwave a viable form for wireless LANs and other forms of open distance communications. Microwave has been used for decades in long-distance transmission services, and by reducing the power it is now being applied for wireless LANs.

There are some problems with using microwave for wireless LANs that put it in conflict with the other established uses of microwave services. One is that the service must be licensed. The other is that there are only 35 microwave frequencies for any spatial area. In some locations, such as a densely populated city, all of the license space may be assigned, leaving none for a wireless LAN. It pays to check on availability of frequencies for microwave before evaluating and considering the use of this technology for a wireless LAN.

Microwave frequencies

The microwave frequencies used for wireless LANs are in the range of 18 to 24 GHz. These frequencies will pass through most things, like tissue and walls. They are capable of a multiplexed traffic load and can be used beyond building limits, if permitted by the license.

Specialized allocations

Microwaves have been used for long-haul, high-speed communications for many years. They formed the original backbone of the long-distance network in the United States. They were also the foundation of companies like Microwave Communications, Inc. (now known as MCI). Long-haul microwave requires territorial licenses and uses high-power microwaves. The wireless LAN microwaves are a specialized part of the microwave spectrum that is licensed for limited distance, lower power operation.

Tuning

Like the infrared LANs, microwave systems must be in a line of sight type of relationship. The tuning is, however less critical than for the optical wave beam, which must hit the optical receiver. The microwave beams spread out in a fan, and as long as a sufficient level of the beam hits the receiver the information transfer can take place reliably.

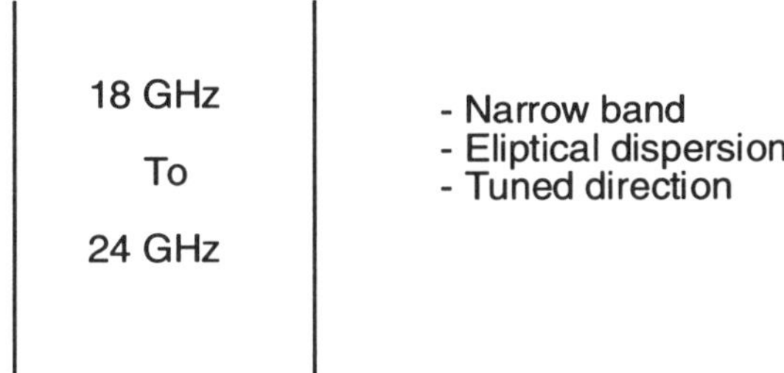

Figure 6.1: Microwave LAN Frequencies

Microwaves are also easier than infrared signals to capture and retransmit in a multi-hop configuration. In most cases the sending station will aim the microwave to a receiver up at the ceiling level of the source room. From here the receiver can be aimed at other receivers at ceiling level around the building. The receiving station will then retransmit the signals to the next level station and create a multihop distribution service.

Distances

The distances for microwave LANs depend on the number of hops between transceivers. The single-hop distance is on the order of 80 to 100 feet. There can be an unlimited number of hops in the overall network as long as the single hops conform to the distance limitations. Long-distance microwave towers can send their waves to around 30 miles, depending on the height of the towers. This is done with high-power microwaves, whereas the wireless local area network systems use far lower power and thus must limit their distance.

Distortion

The dispersion of microwave signals from the transmission antenna is a dispersion pattern that is elliptical in shape. There is some edge distortion in the signal, which increases as the signal travels through the air. This distortion is not a problem, as long as the main signal is received rather than the edge portion. If the relationship angle is off between the transmit and receiving antenna, then distortion or errors in the received signal may occur.   When such errors are detected the signals will have to be retransmitted.

The error correction for most of the microwave systems is built into the transmission hardware. This makes it possible to sample the quality of the transmissions continuously and quickly respond to any faults or needs for retransmission.

Interference

The main interference in microwave signals is due to the existence of stray microwave signals from other licensed users. Many areas of the United States are already saturated with microwave signal licenses and very active patterns of microwaves being beamed through the airspace. These signals, which are usually intermittent in nature, may stray or bounce into the spectrum space of the microwave wireless LANs. This is especially true of higher floors of buildings in major metropolitan city centers.

Figure 6.2: Microwave Interference

Space restrictions

The microwave transmission can travel for a limited distance. It is low in power (below 10 watts) and can bounce off metallic components. The microwave signal is usually aimed toward a ceiling reflector or receiver and then moved by multiple hops along the ceiling space until it is downlinked to the receiving computer network interface unit.

Radiation concerns

Microwaves can be dangerous to humans. Microwave ovens are used to heat items by exciting the molecules within them to create heat. This is done with high-energy microwaves with 500 to 750 watts of energy . The microwaves used for wireless LANs are under 10 watts and are not known to be harmful to humans.

Human flesh will distort the low-power microwaves and interfere with the signals, so there is an equal reason to avoid contact at the human level in addition to the possible effects issue. Moving the microwaves to the ceiling levels of rooms for multihop transmissions further removes the signals from the area occupied by humans (most of the time.)

Microwave pipelines

In their basic form, microwaves are transmitted through a high-speed pipeline that moves data between a transmitter and a receiver using a narrow beam of space and compressed high energy waves. In some situations, such as moving the microwaves through odd spaces, it is possible to use square metal tubes that are hooked together and serve to bounce and guide the microwaves from one location to another. The tubes are called waveguides, and they will contain and direct the microwave signals. This provides two valuable results; the signals are following a known and clean track, and they can be given higher power without problems of interference with humans or other possible signal blockages.

Technically, even with the metal waveguides, the message signal is moving without the benefit of wire. However, the cost of the metal waveguides and their installation may greatly exceed the cost of standard high-capability copper wire. The waveguided microwaves still have the advantage of clean signals, minimal interference, and high security.

CHAPTER 7

Wireless LAN Operations

Wireless LANs support independent, mobile workers. The wireless LANs will need the same operations support and services that are provided for wired LANs. They will also need extra support for their mobility and variable locations of usage. Most of the wireless LAN operations will be automated within the network interface cards and the transceivers. The client/users should not have to perform any special functions or commands to make their mobile client access and use the wireless LAN connection. One of the objectives of wireless LAN operations will be to make them as near those experienced in a wired LAN as possible.

There will be differences in the planning and setup of a wireless LAN due to the sensitivity they have to distance and physical obstructions. There will also have to be support for users who venture beyond the wireless service area and still try to maintain contact with their servers. The cellular systems concept of roaming may have to be implemented in the wireless LAN world.

Planning

The planning of a wireless LAN installation requires more analysis and evaluation than that of a wired LAN. One part of the extended planning is to be certain that a wireless LAN is the correct solution for the users and the application. Another part of planning is to determine that the environment will support a wireless LAN installation and provide an adequate operating environment. Once these feasibility issues are determined, there are additional planning steps for the facility, locations, operational procedures, and the implementation of a wireless LAN.

The first planning step is to determine whether a wireless LAN is a good solution for the application, the users, and the environment. The key factors to be evaluated are the level of mobility needed, the types of data to be transferred, the operational facility space, and the characteristics of the users. The answers that would favor the use of a wireless LAN are:

•   *Mobility of the users*

Untethered movement is required within a physically limited area which could be served with a wireless service.

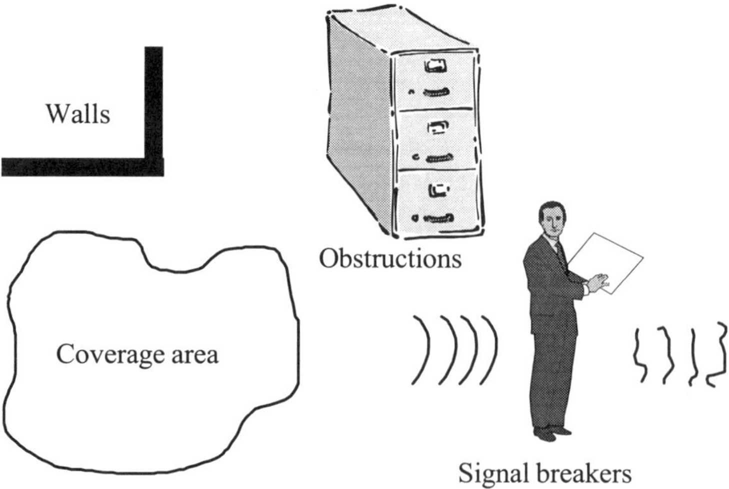

Figure 7.1: User Mobility Limits

- *Data transmitted/received*

Wireless systems work best with short bursts of data being moved. This would include transactions or file reference records and not transfers of large files or big blocks of information.

- *Facility space*

The spatial area should be well defined, contiguous, and self-contained with enough openness to support wireless transmissions without interference or constrictions.

- *User competence*

The users should be computer-adaptable persons who can interact with a network service environment that might have delays or minor disruptions or discontinuity of services.

The second planning step after the successful determination of the feasibility of using a wireless LAN is to evaluate the facility space and do some preliminary layout of the wireless LAN environment and service locations. It is useful to obtain architectural layout drawings of the facility area where the wireless LAN will operate. These diagrams should show the major structural components, such as walls and supporting columns, and the temporary additions such as partitions, partial walls, and false ceilings. It is often useful to have access to diagrams that show the power, heat, ventilation, and other utility services in the area including telephone, wired LANs, and other communications services. If the facility already has wired LANs installed, these diagrams may already exist with the LAN components drawn in place on the diagrams. Always start with the most complete reference diagrams possible. Ones with the current LAN layout are best because the wireless LANs will likely be connecting to the wired world at one or more places.

If the wireless system is to be the first installed LAN and/or there are no adequate facility diagrams available, then it is recommended that someone make rough sketches of the areas to be covered and use these as the planning base for the wireless system. The sketches do not have to be architecturally perfect, but they should show the major spatial areas with measurements and indicate any fixed elements that the wireless system will have to interact with in its transmission and receiving area. The fixed elements would include:

- Walls

- Structural columnsDoors and windows

- Interference sources

- Major equipment installations

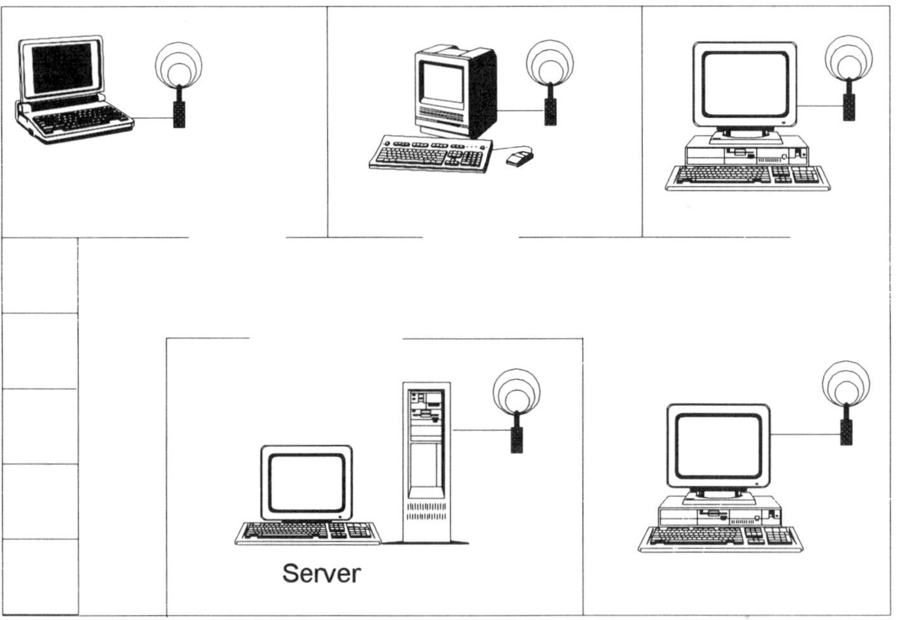

Figure 7.2: Typical Space Layout Facility Diagram

Once the diagrams are available, the next step is to plan the usage or wireless layout of the spatial area. The determination of logical transmit and receive points to be within the distance limits of the technology and easily accessible to the users will take some amount of trial-and-error positioning and evaluation of the possible operations of the wireless LAN. The key factors in the positioning decisions are:

- *Clear access*

Can the points be reached and used without having any restrictive building components in the way between the transmit and receive facilities?

- *Logical user interfacing*

Can the access points be conveniently used by the application users without difficulty or blockage of their efforts to communicate using the wireless technology?

- *Locational movements*

Can the users move comfortably through the environment and still be able to be in send/receive contact with the wireless LAN?

- *Possible voids in service*

Are there any places where there will be likely difficulty or possible inability to maintain services to the wireless LAN? Known as voids, these areas should be minimal to nonexistent. If they are present they should be in places where it is unlikely to be necessary to transmit or where the voids will be a minimal problem. Too many likely voids may invalidate the idea of using a wireless LAN.

- *Roaming*

Are there places where one wireless LAN can detect and hand off services to another LAN? Roaming connectors should be placed where they have the greatest area of coverage and can efficiently move users on the limits of one LAN to another without loss or disruption.

- *Natural warning locations*

Does the spatial area have natural borders where warnings can be posted or issued to inform the users they are at the boundary of the wireless LAN's service coverage?

In some situations it may not be entirely possible to predetermine the feasibility of a wireless LAN. In such cases a trial evaluation implementation may be needed. By setting up a small wireless LAN in the physical environment, it should be possible to confirm or reject the potential for a successful full-scale operation.

Setup

The setup of a wireless LAN system involves the physical delivery and establishment of the send and receive units. This may include facility preparation and organization, determining the location of transmit and receive equipment, and review of expected applications for the wireless system.

Setup also involves defining the server units and the client units in terms of their capabilities, services, identity, and locations on the network. The interfacing to existing computer resources and wired LANs will also need to be evaluated. Such systems will also need to be reviewed for the adequacy of their resources to service the loads from the wireless LAN world. The added loads from the wireless world could have major impacts on the use of available memory and disk space.

One way to evaluate the possible loads that wireless connections will add to the existing information resources is to identify the types of transactions that will be used in the wireless information connection. For example, a bar code data collection system will be likely to make a request for product or work unit information approximately 20 to 30 times per hour. If each request is 50 characters long and the return product or work unit files average 3000 characters, then the overall transaction flows can be defined and estimated. It is typical that mobile wireless users will have more outbound (return) data traffic than inbound (input) data.

By working out the bidirectional flows of information, the designers can determine the additive loads from wireless users and then evaluate whether the existing resources are adequate to handle the work. If they are not, they will need to be upgraded, split, or reconfigured so they can support the new total work effort.

Once the physical and logical elements have been determined to be adequate for the wireless LAN installation, the setup plan is complete and the physical installation can begin. The information for the setup should be documented and made a part of the systems definition and management file.

Installation

The installation process for a wireless LAN varies slightly with the chosen wireless technology. The common elements include:

•   *Install the transceiver units*

Installation may consist of installing a network interface card into the computer unit and coupling wires from the card to the transceiver unit. The transceiver units will need to be aligned and tested according to the instructions provided by the vendor. Usual placement involves keeping the antenna in a higher space that is clear of people and physical interference.

•   *Tune the send and receive elements of the systems*

The tuning process consists of selecting the starting frequency levels and running some self-tests to see if the units can establish and maintain contact with one another. Adjustment of the tuning alignment consists of moving the locations and aiming the transmitters and receivers connected to the various computer units. The tests would then be rerun to determine whether any improvements were generated. Other tuning can be done with a multifrequency signal strength meter. This unit is placed in various locations within the intended signal space to measure the level of received signal when the test units are transmitting.

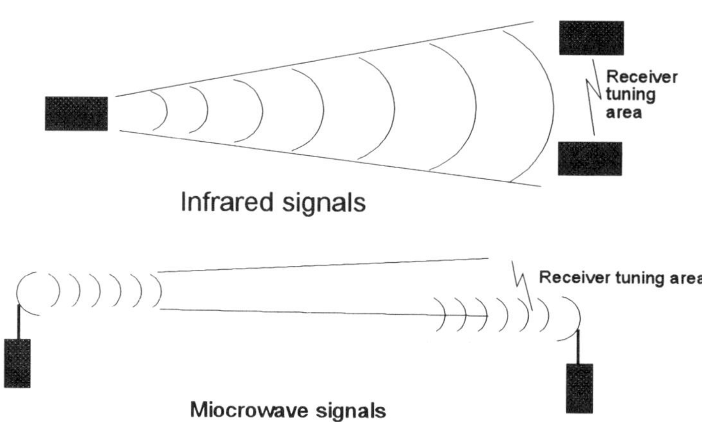

Figure 7.3: Tuning Wireless Transceivers

- *Load the software into the clients and servers*

Once the transceivers are in reasonable tune and the units can transmit and receive to/from one another, it then becomes time to set up the software that will connect the applications and the user to the wireless LAN world. The loaded software is usually a vendor-developed version of a mini network operating system. This software needs to trap data operations that must move over the wireless network and develop the proper data and messaging formats to be used for communication over the wireless LAN. This software will require resident space on the client and the server systems. It will trap user/application requests for remote services and format these requests into forms that can be recognized and serviced by the wireless LAN.

- *Test the components for proper operation*

This level of testing will involve both the applications and the wireless network components. The application level on the client should set up a request or an information transfer that is intended for a remote server reachable over the wireless LAN. The test should involve observing the client and server relationship over the wireless LAN. The steps, actions, and timings of the client generating the wireless LAN transaction and the server responding to it would be tracked and evaluated. Once the tests are functionally complete and deemed correct, the next level of testing will come when the system is operational and multiple transaction loads begin to challenge the system's performance.

- *Test the connectivity components*

The test of the connectivity components will validate the ability of the wireless LAN to sustain reasonable throughput under varying situations and conditions. Tests for loads, continuity of service, throughput timings, and other performance characteristics should be run and evaluated. If the performance is not deemed adequate, then adjustments should be made to priority schemes, network segments, reliability characteristics, speed increments, and/or other adjustment factors.

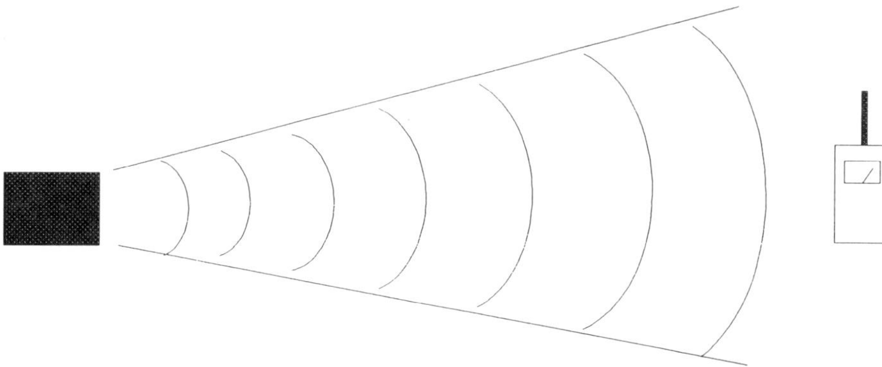

Figure 7.4: Testing of Transceiver Signal Strength

- *Train the users*

Most of the operations of a wireless LAN should be transparent to the users. However, there are several unique operations and steps that the user will need to know and support to make the wireless LAN a success. One area of training is in the setup and maintenance of the physical arrangement of the wireless units. The placement of antennas, transceivers, and other coupling components is critical. The locations for the units that are known to work best should be identified by the users. In the event of decoupling of the wireless LAN, the user will need to know what to do to attempt to reconnect and initialize the services of the wireless LAN.

- *Document the system*

The operation of the wireless LAN and any work the user will need to do to keep it running will need to be put into written form. This documentation could be provided in an on-line file format, an interactive help service, and hard copy. The material should be clear in format and well indexed so that users can find the documentation portion that is relevant to their needs.

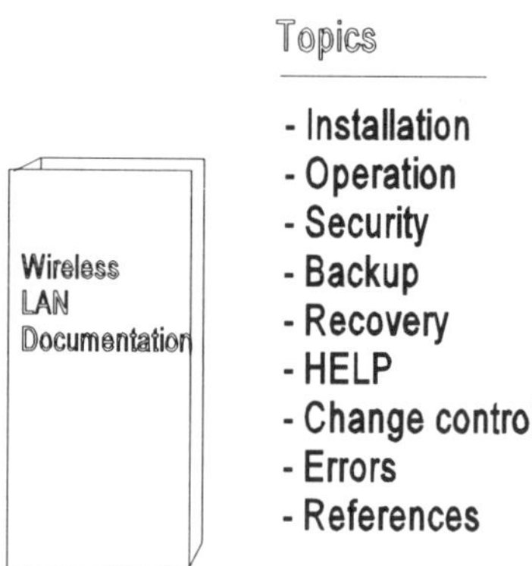

Figure 7.5: Sample Wireless System Documentation Contents

- *Set up the operations support services*

The wireless LAN installation process is a short one-time effort for each node. A few hours are all that is required for the opening, connecting, and testing of the wireless LAN. Most of the time will be taken up with the installation and testing of the software components. This may require reallocation of memory space, modification of user files and systems specifications, and adjustment of user applications.

- *Final operational testing*

The total installation must be tested under realistic operating conditions. Sometimes user applications programs will not run as the network modules have taken up resident memory and prevent certain user applications from loading into their normally expected areas. Considerable adjustment and shoehorning may be needed to make the system properly operate all user applications and services.

Merging wireless and wired LANs

Very few wireless LANs operate as stand-alone entities. Most of them are the local user front end to a fixed wired network environment. The wireless LAN may support the mobile workers, but it will at some point connect to a wired LAN of the organization. Wired LANs have been the traditional form of LAN and make up most of the connectivity between the offices and operational facilities of an organization. The wireless LANs are more recent introductions that are used to service special classes of users and provide them with capability for both the wireless and the wired world of data and information.

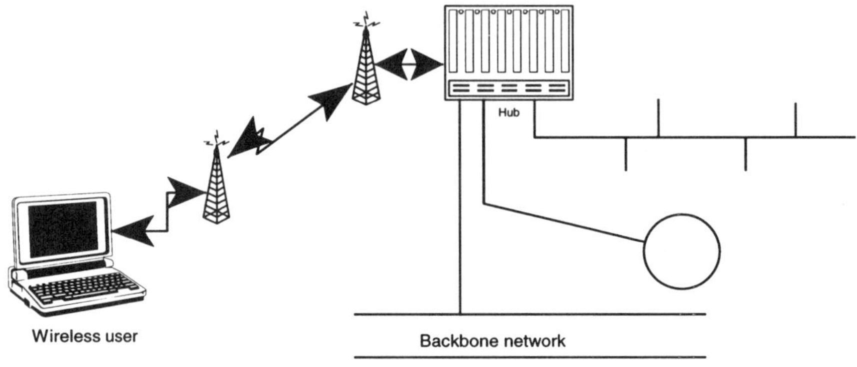

Figure 7.6: Typical Wireless-to-Wired Configuration

The merging of the wireless and the wired worlds will logically happen at one or more server nodes within the networks. Either special connector nodes or extensions of a working server will be used to couple the wireless and the wired worlds. From this coupling information can flow bi-directionally from the users to the server and from the server to the users. The traffic will flow over whichever technologies are used to connect the units together.

Software interfaces

The software interfaces between wireless and wired LANs will require a more complex and difficult process than just moving a signal from one medium to another. The software interface will need to recognize messages, transfer requests, packets, error checking, security, and other operational elements and keep these factors under proper control for each of the environments and the users.

Software for the wireless LAN services will be loaded into the wireless clients and into the wireless servers. Some of the software can be extensions of the network operating environment. Other software will be service modules for wireless protocols and messages. Still other software will be application based.

The application-based software will connect the end users with the processes they will want to exercise in a wireless world. For example, PDA users could tap their screen to bring up a database access query form, which they would fill out and then release to their wireless communications manager. The wireless management software would frame and transmit the message over the wireless medium and then wait for a reply from the wireless server. The wireless server would receive the database query and then determine where and how to provide the requested answer. Once found (which could involve multiple server searches in the wireless and/or wired world), the information record would be formatted and transmitted back to the requesting node.

In the case of wireless bridges, the transmission management software would determine that they had traffic for the wireless bridge and direct it to the appropriate wireless unit. Once it was received, the wireless bridge would execute the bridge interface software and move the data to the receiving location. The actual application and servicing of the messages would be beyond the concern of the wireless bridge software. All the wireless bridge software would do is move presented data to the receiving location. In the reverse situation, the wireless bridge would receive inbound traffic and upload it (in the proper format) onto the internal LAN for further processing.

Most of the software interfacing should come together in the use of a common operating system on both environments. This would allow for the building of message process management at the operating level. To cross between two different operating systems, environments would most likely use the upper level control facilities of the OSI model, which was built for a more monolithic structure.

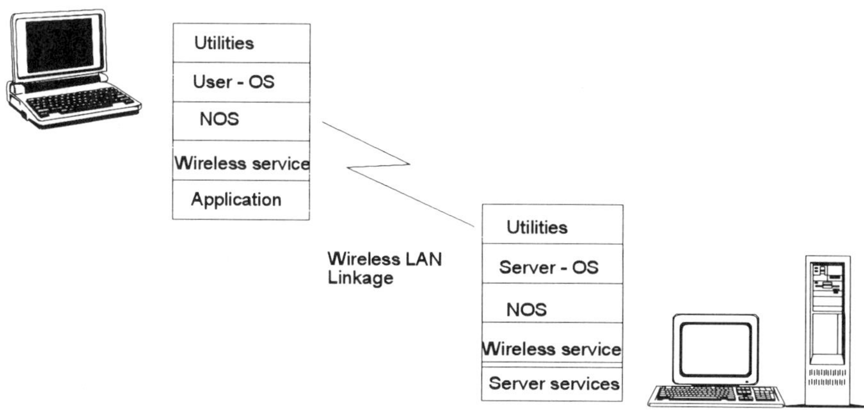

Figure 7.7: Typical Software Configuration

There are plans and development work under way to support mobile clients within Microsoft Windows NT Servers and within Novell NetWare 4.1. The IBM OS/2 LAN Manager and the Banyan VINES systems will also follow suit. The new release of Windows 95 also provides support for wireless networking, remote users, roaming, infrared, and other components of the wireless LAN world. The more the wireless world becomes a supported option within other standardized worlds, the more wireless will become an accepted concept for user applications.

Security

The security of messages and data that move from a relatively protected wire-based environment into and out of the wireless world will be a continuous concern. Some experts remind us that the security of the wire-based environment is not as good as may be presumed. However, the thought of important messages and data flying around in an unstructured environment subject to receiving interception without detection is a significant cause for concern and worry in the wireless world.

In reality, there are several parts of the design of the wireless technologies that actually increase the security of the environment rather than reduce it. The ability of spread-spectrum radio signals to perform frequency hopping as they travel from source to destination makes it inherently difficult to identify and select a signal from the environment that is whole and meaningful, unless one were tapping the transceiving server that was doing the frequency hopping.

For those who still worry (or are paid to worry about LAN and data security), there are several technological additions that can be made to the wireless LAN world to improve the security issues significantly. One of these is to encrypt and decrypt all sensitive data. This is often done in the wired world, and the same algorithm and mechanics can be applied in the wireless world. Another approach is to log and trap the transmission activities of the wireless world and maintain high vigilance for unauthorized users and/or transceivers. Physical security, user identification security and facilities surveillance will all help to limit the possibility of unauthorized intrusion into the wireless LAN world.

Most of the security management processing for wireless LANs is done at the software levels. Encryption, decryption, audit trailing, etc. are all done via software modules. The mechanics of the process are the same in the wireless world as in the wired one. However, the software will need to deal with wireless communications sequences and be able to become part of the wireless operations process. As in the wired world, most of these security functions can be made resident options within the network operating system layer.

For example, within the OS/2 Warp 3 edition there are security functions and choices that can be applied to any of the supported network forms. If the OS/2 product is used in a wireless LAN setting, then the specification of OS/2 is that it is to use its wireless LAN capabilities. The choice of security processing and the levels of security to be implemented are those that come with OS/2. They would be the same choices whether the system uses wire, wireless, or a combination.

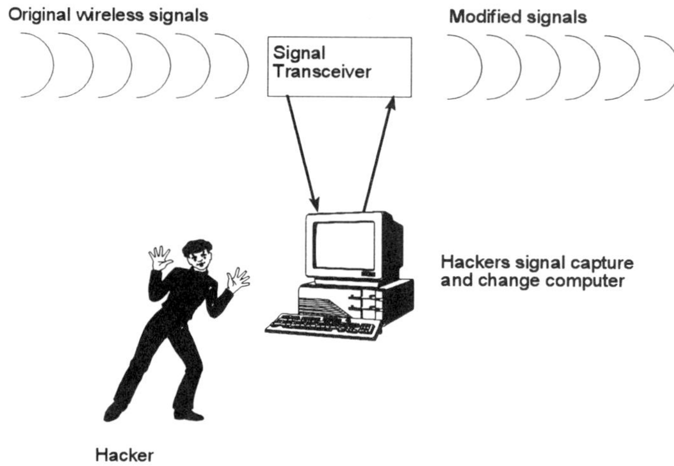

Figure 7.8: Wireless Hacker-Eavesdropper in Operation

The security of information coming in over longer distance wireless LANs or from facility remote users will require a higher level of security than is applied to users who are within the physical facility environment. User identifications, call-backs, and higher order encryption and decryption processes may be appropriate.

File and data management

The passing of files and data between the wired and wireless environments is one of the major traffic levels between these two environments. Remote wireless users will want access to files and records from databases that are maintained in the wired world. They will want to interact with the wired databases in order to access and process changes to the data components.

The wireless-to-wired data connection will require inclusion in the overall data management scheme of the information applications. The data from the wireless world will need to pass security and content scrutiny that is somewhat tighter than that in the wired world. The key issue here is that data management in the wireless world will involve no less than that applied to users operating in the wired world.

Good file and data management will mean that all file activities should be framed into a structured transaction that can be identified as to form, source, legitimacy, content, and operation. A change request to a data field should be able to pass a set of logical tests that are based on the application and the data components, not whether it is coming from a wireless or a wired world.

File and data management issues involve systems-level design decision and should not be influenced by the form of communications coupling being used. If the wireless format raises special concerns for the data management, they should have been present for the other forms of interfacing. However, if the proposed use of wireless LANs increases concern for data management, then it should be applauded as a positive factor in raising consciousness of this issue.

Message transfers

The wireless LAN will also be used to move messages between remote or mobile users (clients) and the enterprise-level networks and servers. The messages could be transactions packages, notices, electronic mail, groupware communications, etc. The message traffic is more flexible and variable than the data transaction traffic. It can be scheduled over a more flexible time environment and can be managed in a more subtle and consistent format.

Intelligent agents can be used within the wireless LAN to help with the data transaction traffic. An agent is an intelligent program that has capabilities for recognizing conditions and initiating its own sequence of operations or calling other programs into execution. Agents can be used to unclog data communications traffic, direct database searches, move information from one location to another, and perform numerous other tasks without user or programmed intervention.

Agents can detect the arrival of wireless LAN messages and then perform the translation of the wireless message into a wired format. They can interpret user requests and initiate the proper response programming. Agents need to receive a message and be able to interpret it. Once they have decoded the request, they will determine what steps to take to make the required response.

Agents can be used to support operational services, user requests, specific processes such as database lookups, etc. Given the high level of user connectivity in the wireless LAN world, agents can be a good way to provide user transparency and maintain good operating procedures.

Message transfers are a good candidate for the use of agents. The wireless message traffic can be slotted into available network space, prioritized into different categories of service, and generally handled in a more orderly manner than on-demand interactive data transfers and queries. This does not mean that the messages can be forgotten or delayed randomly. It means the network can manage the message rather than simply waiting for interactive user demand. Intelligent agents can supply the oversight and direction to see that the requested services from the application or the user are provided in a responsive and controlled manner.

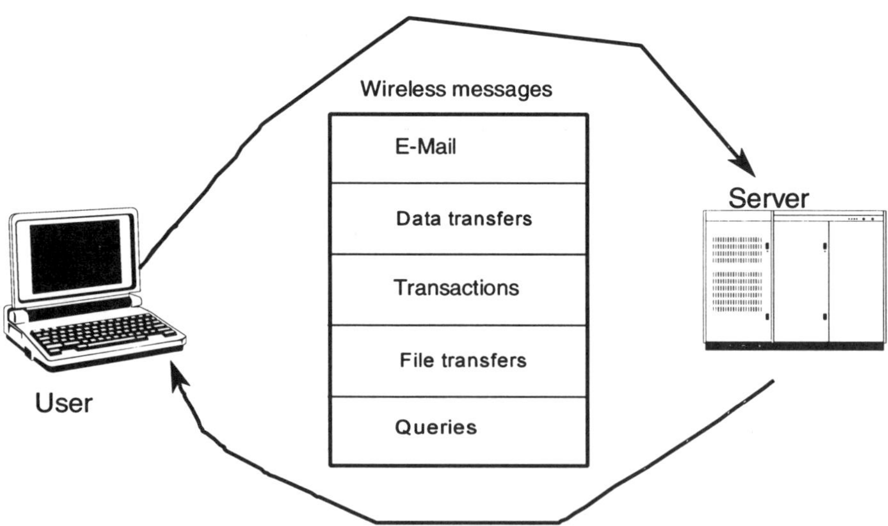

Figure 7.9: Wireless Messaging

As most wireless networks will be a mix of message and data traffic, the mix will determine the management orientation. A wireless network in a warehouse that is handling logistical goods movement will be 90% transaction handling and only 10% message traffic. Hence availability, service continuity, reliability, etc. will be very high. In an administrative work group setting the wireless LAN may handle 90% message traffic and only 10% transactions. This would significantly reduce the loads and service provisioning of the wireless environment.

Wireless LAN Operations in Perspective

Wireless LANs are still LANs. They are intended to move data, information, and messages over a limited distance spectrum, with close connection to the end users. Their operations will often be just one link in a long chain of events. The wireless LAN must be a cooperator with other forms of communication. It must neither dominate, dictate, nor control the movement of information between the various parties. The operations should be integral to the mechanics of the information transfers. Where possible, intelligent agents can perform many of the functions of the wireless data movement and coupling to other network services.

CHAPTER 8

Wireless LAN Applications

Wireless LANs can support a number of applications in a wide variety of organizations and industries. Most of the applications focus on the mobility of the user or the temporariness of the application. When the application fits the characteristics provided by the wireless LAN, then the system can be successful as long as it fits within the physical limits of the wireless technology.

Mobile operations

The wireless world is oriented to supporting mobility. If the user (client) did not move, then the tried and true wired approach would be the preferred choice. But as workers in many situations are mobile by nature and/or would like to become more mobile by choice, the wireless world provides the technological linkages to the established data and information services available in the wired world.

Mobile operations can be conceived in many different forms and implementations. A telecommuter is a mobile worker, as is a traveling salesperson. In the wireless LAN context a mobile worker has to be within the physical area of contact that can be maintained by a wireless LAN. This can include the inner space of a building, a local campus or a multiple facility environment that is within a line-of-sight connectivity zone.

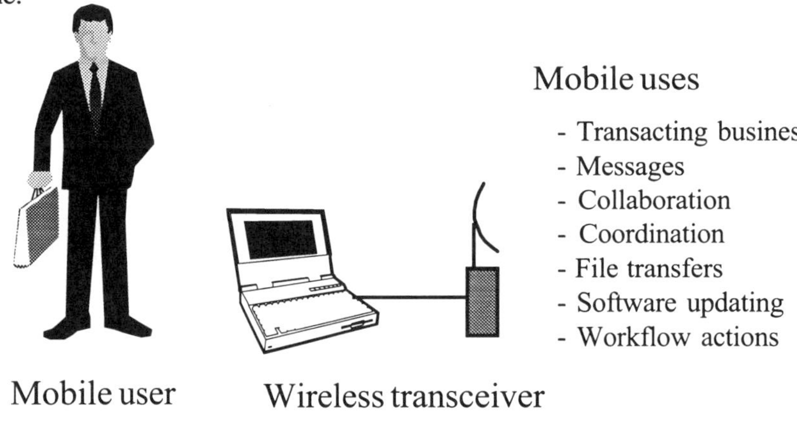

Figure 8.1: The Mobile Worker in Action

Mobile operations usually mean that the individual users cannot easily be connected to or use the traditional form of wired desktop-supporting LAN. The mobile user is moving about too much to be able to connect via a fixed-location client station. As mobile users move about in their environment, it is important to take note of the fact that the level of mobility and the spatial areas covered by today's wireless LANs are also limited and contained. The tighter the user's roaming space, the better for most wireless LANs.

Infrastructure based

The infrastructure consideration for mobile workers is that they operate within a fairly tightly constrained area of the organizational facility. Wireless LANs are meant to be limited-distance support products. They cannot support open and un-limited roaming of the users. The supported infrastructure is more oriented to a department area or a small section of a facility, not unlimited space or the total facility unless it is very small.

The infrastructure limits of wireless LANs are also difficult to live with in many organizations. The boundaries are ones of going beyond the limit of the communications technologies; they are not hard and fast limits like the walls of a department or building. What happens is that the user simply fades out of contact with the wireless world or may sense a cutoff type of disruption in services as they reach beyond the established limits. Due to interference and spatial considerations, the break in service may vary by location, direction, time of day, and other random factors. In other words, what may have worked yesterday, may not work today.

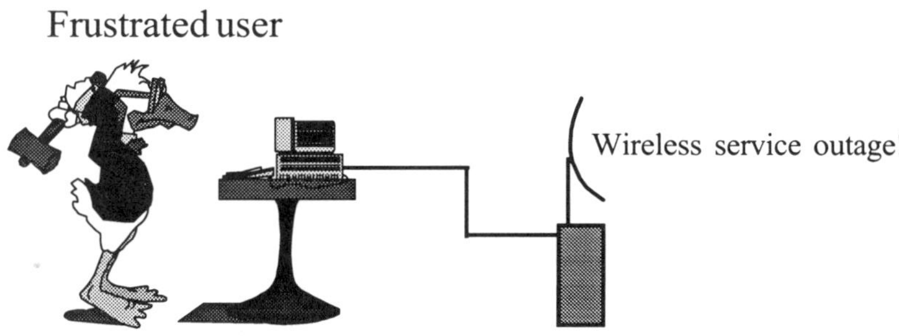

Figure 8.2: Frustrated Wireless User Being Cut off

Flexible connectivity

Flexible connectivity means that the users can choose when and how they want to be connected to the wireless LAN services. Some will choose to remain in such a contact state on a continuous basis, and others will make the connections on an occasional as-needed basis. The key concept is the ability to obtain services when and where they are needed without regard to the specific positioning or location of the servicing units.

Flexible connectivity also means the ability to connect to a broad range of services and/or to different networks as the need or location may dictate. The wireless LANs can support flexible connectivity by providing an automatic recognition of roaming users and the ability to hop across various networks (wired and wireless) to obtain information and services to which the user is entitled.

Spatial limitations

The wireless LAN always comes with spatial limitations and restrictions. The dimensions of service that wireless LANs can provide are limited by signal strength, facilities limitations, and user position and mobility. For example, an office environment such as a school can provide better facilities for wireless operation than an industrial plant or a hospital. Both of the latter institutions are likely to have machinery and systems that will impose restrictions on and operational interference with the successful operation of wireless LANs.

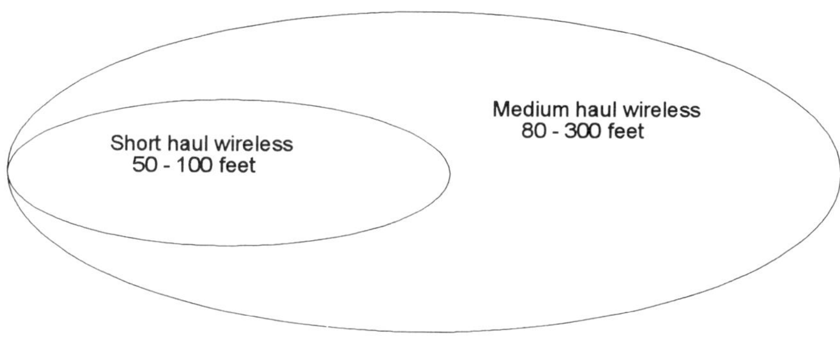

Figure 8.3: Comparison of Wireless LAN Distance Limits

The spatial limitations do vary with the specific wireless LAN technology and the actual physical facilities in which it will be used. Open areas can enhance some wireless technologies and restrict others. Walls, ceilings, and other physical barriers can be a hindrance to some systems and a boon to others. One needs to consider the application, the space, and the technology options before specifying which wireless LAN might work in the setting.

Connectivity management

Connectivity management is a critical component of the provisioning of the on-going services of a wireless LAN. As the user location can change, the connectivity management process must not only verify and control the security of who is using the wireless system but also be able to locate and track the users so they can receive the information requested or addressed to them.

Connectivity management will also oversee the interfacing of messages and requested services that will cross over from the wireless LAN to the wired LAN environment. Requests from either direction will need to be captured, logged, validated, and managed to successful completion. In the event of a disruption or incompleteness of requested services, connectivity management will need to intervene and reestablish the services.

Connectivity management is a continuous function that must operate on good transactions and bad ones. The continuity of services to the user via a wired or wireless LAN is equally important and should be transparent to the user. This is accomplished by a combination of hardware and software facilities that constantly monitor and validate the transmissions and messages sent over the wireless LANs.

Special services

The special services to be provided via a wireless LAN are mostly those dealing with variable locations. Access to various software, databases, and equipment resources via the wireless LAN may require special interfacing devices, multiple network paths, and/or unique security mechanisms. For example, mobile users may have a pager-type device that alerts them to the arrival of a wireless LAN message or transaction that is intended for their receipt and handling. Another example is a large, easily read display panel mounted on a forklift truck that alerts the driver to a request for movement to a particular location in a warehouse. The forklift could be coupled to a wireless locator network that tracks its whereabouts within the warehouse using a form of micro cellular wireless technology. Other special services involve the use of wireless transceivers to various service request devices such as printers, displays and other presentation devices. Other wireless service connections can be made to local and personal file devices located somewhere within the network architecture.

Flexible workers

Many work situations require the worker to move around in the work environment, making it hard to make use of a standard computer. Flexible workers can be serviced by portable computer units that are connected to a wireless LAN to give them communications and server access.

Health care

Health care work is one typical form of flexible work that can benefit from wireless LANs. The provision of health care services is centered around the interface of the patient and the health care provider. As medical providers usally go to many patients, it would be difficult, expensive, and inconvenient for them to constantly shift from one computer to another. It would also be difficult to provide a tethered system, because the care givers often have to move long distances between patients. Thus a wireless LAN can be one answer for a health care/hospital environment.

By using small portable computers such as personal digital assistants (PDAs), palmtop computers, or pen tablet computers the care giver has portability and a sufficient system platform to input and output the information needed to perform and deliver the patient services. The connection over the wireless LAN will take the individual user units to a local server where patient data and service information is stored and retrieved. Input transactions can be validated at this local server and the data passed to other LANs and servers for processing and formal file updating.

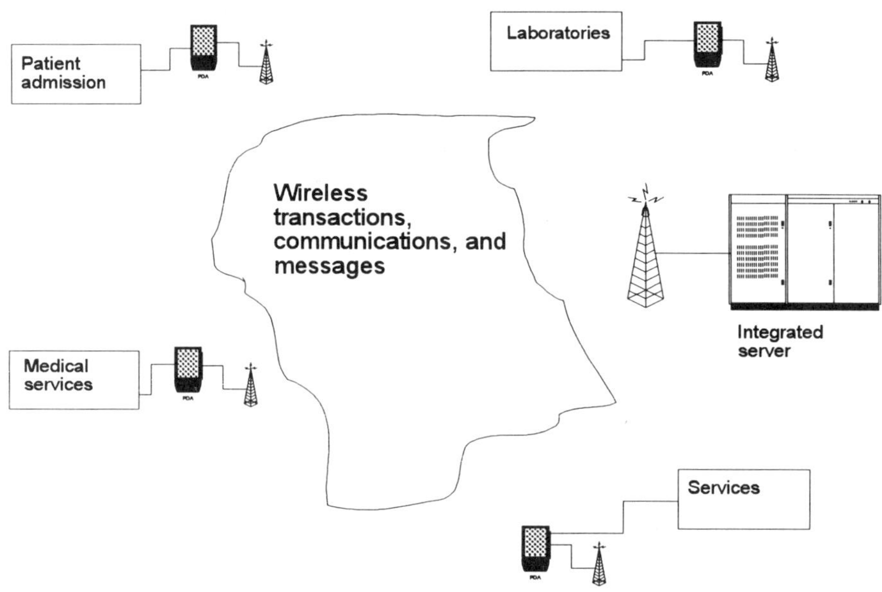

Figure 8.4: Hospital Data Flow Diagram

When care givers arrive for their rounds they will take a portable computing unit from a service rack and initialize their identification and security codes. The unit will then extract the patients to be seen and the services to be performed from the local database and provide the list to the memory of the portable unit. As a care giver moves to a patient's location he or she would select the patient's name or identity number from the list and the wireless LAN will request and return the appropriate details on the patient to the portable unit, where the user can view, list, browse, and interact with the patient's file.

As the care givers perform their services they can have check-off lists presented on the portable unit to indicate conditions and prescribed services. They can also enter notes, recommendations, reminders, etc. to the patient's record. When they complete contact with the patient, they can finalize the data recording and dispatch the revised file back to the local server for updating of the patient's records. Any new service initiations (medication, tests, etc.) would create transactions at the server which would be dispatched to other servers in the appropriate health care service systems.

In this application the wireless LAN provides support for the mobility of the health care provider and the capture of information at the point of origin. It also distributes appropriate information from networked servers into the hands of the health care provider at the time it is needed. Because of the volume of patient information involved and the frequent necessity of responding to unexpected events in the health care environment, the wireless connectivity to the LAN provides more complete and navigatable information access than if the data were simply down and uploaded from a portable PDA device.

Health care administrators

Most of the focus of wireless services in health care is aimed at the care givers. The administrative side can also benefit from some process reengineering and the use of wireless LAN technology. Health care services are known for their heavy loads of paperwork and numerous lines and administrative forms that must be completed before the medical services are provided. This process can be difficult and stressful for someone who is waiting for the medical procedures (already a stress-inducing situation).

Several hospitals are changing their admissions process by using laptop computers and wireless LANs. When planned patients arrive at the hospital they are identified and immediately escorted to their rooms, introduced to the area staff, and invited to become comfortable with their surroundings. A health care admissions representative arrives at the patient's room equipped with a laptop computer coupled to the area wireless LAN. The representative will take all of the data and patient information interactively and complete the form using the computer. Access to files and records and updating of the results to the departmental computers are done over the wireless LAN link.

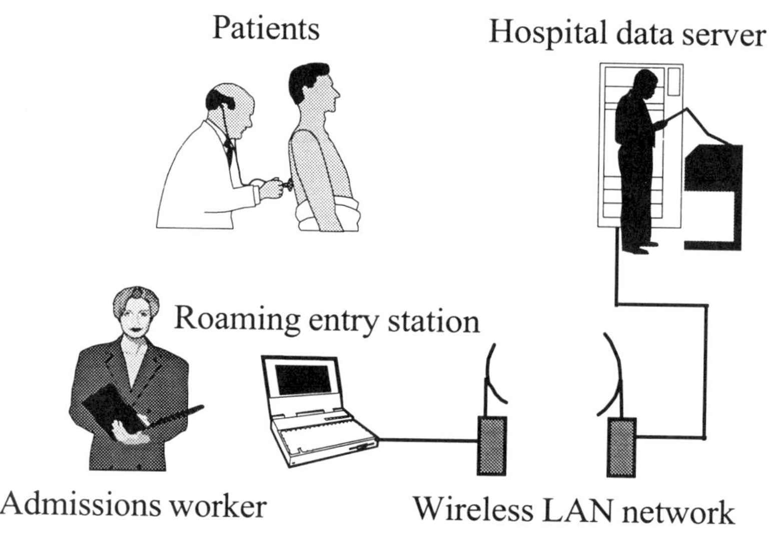

Figure 8.5: Health Care Admissions with Wireless Computing

The result is a calmer patient who does not have to worry about the administrative hassle. The health care administrators benefit as well, because they can schedule their visits to in-place patients and do not have to deal with lines of less than happy clients.

Inspectors

Inspectors whose job it is to investigate and evaluate the condition or quality of a product or system often need access to large files of data and information and in turn produce data that must be shared and updated in product repositories. The inspection process may involve a piece of complex equipment on an assembly line (an airplane, locomotive, machinery, etc.), a system ready for use in a complex and difficult operation (the NASA space shuttle), an aircraft in for repair or FAA recertification, or the status of work on a major construction project site.

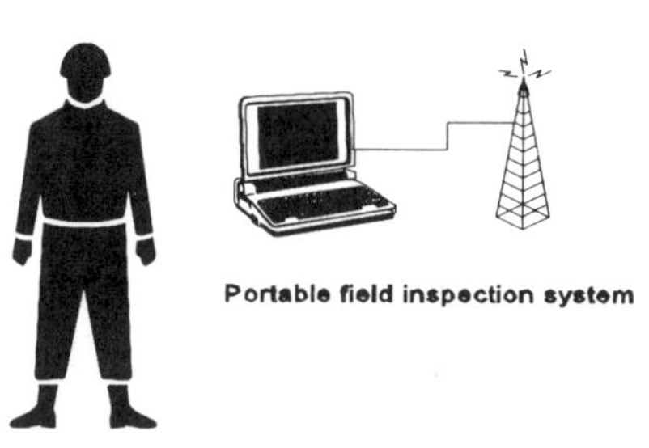

Figure 8.6: Inspection Data Collection Process

Key to all of these inspection activities is that the inspector must go to the location of the inspection and away from the desk environment. Inspectors are also often in an environment which does not have easily accessible computer systems. They also have to move around in a spatial area to conduct their inspection activities.

Where the inspectors need access to data about the product, previous inspections, tests, status information, etc. a wireless LAN can give them access to the local and remote information server worlds.

Service agents

Service agents can be employed in a broad number of situations. A service agent may be a roving repair person inside an organization or a waitperson in a restaurant who needs to be in contact with the kitchen and food preparation areas.

Service agents often need to transfer information about the service situation, such as a trouble ticket or a patron's order, and receive information about the history or status of the situation. The information transfers are bursty and time or event triggered.

Service agents are also found in roving stations within a larger service organization or institution. Many airlines are now using wireless LANs for their roving customer service agents who support personal customer service within an airport terminal. The airport service agents need to move between gates, support the location of passengers, and interact with other airline services and support activities. The wireless LANs provide the flexible connections between the service agents and the operational databases of the airline.

Among the first service firms to use the wireless LAN concept were the rental car firms. They can equip their rental agents with portable computer stations and have the agents meet the customer at the car as they rent or return the vehicle. The unit can check out and check in the car by communicating to a local server over the wireless LAN. The customer receives quicker service and the rental agents can tend to more customers. The costs of equipment are about equal to those of the fixed counter systems, with savings in building space and facilities. Only inclement weather dampens enthusiasm for this type of service solution.

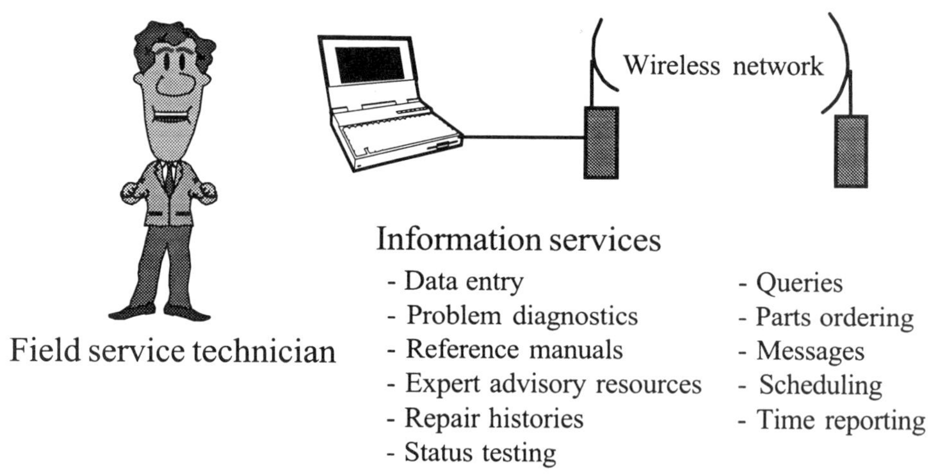

Figure 8.7: Field Service Wireless Perations

Teachers

Teachers in a public or private educational system (K-12) are logical candidates for the use of a wireless LAN. The campus facilities of most schools lend themselves to a zone type of wireless LAN setup and the teachers are highly mobile as they bring their teaching efforts to various groups of students.

The use of a wireless LAN would provide the teachers access to student information, grade books, teaching resources, lesson plans, attendance, messaging and a host of other information systems and services. It would allow the teachers access to data at any location within the teaching campus and the ability to take their client workstation (most likely a notebook computer or a personal digital assistant) to any location where they would be dealing with the teaching process or their administrative duties.

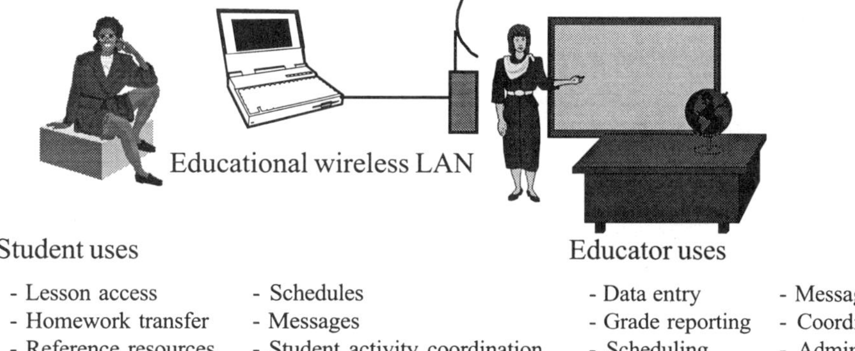

Figure 8.8: Educational Wireless System

As teachers are a highly mobile workforce yet stay within a contained campus environment, the wireless LAN would provide direct support for most of their data processing needs. Additional services such as access to the Internet and other teaching services would also be available through the wireless LAN.

Pit brokers

Pit brokers are individuals who operates in an auction center type of environment where they are buying and selling some item at a bartered price. This would include stock markets, commodity markets, and animal auctions. All of these operations are highly mobile and transaction oriented. They also operate within a confined spatial area. This makes them ideal for the use of wireless LAN technology.

The pit brokers would utilize handheld computer units on which they could write or tick off choices, volumes, and prices as well as indicate customer identifications. Once they completed the entry of a transaction, the unit would transmit it to the local server. There will be numerous units operating in the same time framework, so capacity and channel capability would be important. In addition, the handheld unit would receive buy-sell orders from the back-office personnel to the floor (pit) broker. He or she would have to process the order with the other brokers and record the final result of the negotiations.

The wireless LAN provides the ability to maximize the flexibility and mobility of the pit broker while completing the data on the trades as they occur, Time and mobility are paramount in the brokering application. The wireless LAN provides both.

Temporary situations

Temporary work situations are excellent candidates for the use of wireless LANs. The systems can be set up quickly in most compact locations, operated while needed, and then taken out when the temporary situation is over. Typical temporary situations include:

*   Sporting events

*   Entertainment programs

*   Political campaigns

*   Fund raising campaigns

- Special events

- Training camps

- Seasonal events

- Sales campaigns

Ad hoc connectivity

Ad hoc connectivity usually means temporary but on-demand connections to the LAN services world. This type of wireless LAN would require rapid availability, but erratic and low levels of use when there was no ad hoc condition.

Typical ad hoc activities might include:

- Security operations

- Special sales

- Management events

- Emergency situations

- Conventions

The ad hoc application of wireless LANs would require the ability to orchestrate a rapid connection to the existing wired LANs. This would include a physical transceiving station to make the connection to the wired world and the ability to test and validate the connection. It will also be necessary to identify and validate the users who will be signing onto the system via the ad hoc wireless units.

The ad hoc applications can literally be here today and gone tomorrow. The ad hoc setup and the necessary equipment would probably be stored as a library unit, so it could be quickly acquired and implemented. Pretesting and trial runs at setting the ad hoc services up and using them would be mandatory to prevent any glitches from developing when a real ad hoc use was being implemented.

Event support

Event support services from a wireless LAN might include:

- Registration

- New notices

- E-mail bulletin boards

- Accounting

- Scheduling

- Summarizations

- Other services

Event support from wireless LANs would differ from ad hoc situations in that the event will be planned over a longer period of time and there will be more time to orchestrate the system and services and to set up and test them. In addition many of the services for an event may need to be specially programmed, whereas the ad hoc uses will be more predefined or canned.

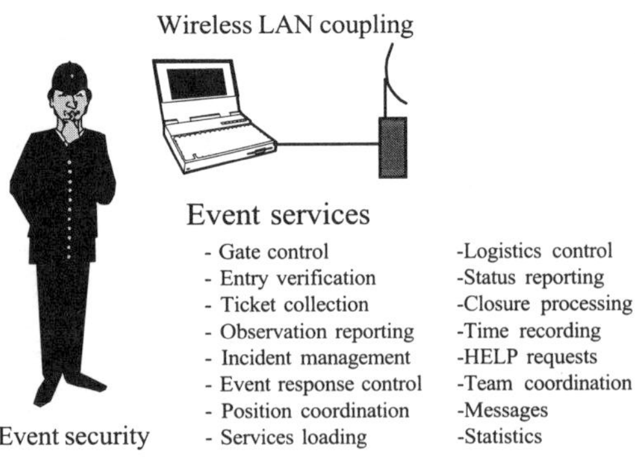

Figure 8.9: Wireless LAN Support of an Event

General services

The provisioning of general services for an event or temporary situation would involve the coordination of mobile workers performing general support activities. An example would be the cleanup work after a major sporting event. The cleanup crews could use portable client computers to coordinate their locations, work efforts, needs, and status. As they move to various cleanup functions, the wireless LAN would be used to update progress and determine work orders for further efforts.

The coordination of support services such as trash pickup and/or delivery of support resources can be done via the wireless LAN. Services such as time keeping, consumable supply usage, and messages on other needs can all be supported via wireless LAN systems.

Emergency centers

Emergency centers have a tendency to be here today and gone tomorrow. They are often highly localized, command center–oriented situations that need coordination and control of diverse resources working to correct a defied emergency situation. A fire, flood, disaster, accident, or other catastrophe would qualify as an emergency service situation.

The emergency itself could last from a few hours to many weeks. The efforts could involve rescue efforts, medical attention, property modification, relocation support, service provisioning, and emergency restoration services. The key in servicing emergency centers with information is the speed of setup and the flexibility to maintain information services between a diverse set of mobile workers. The unpredictability of the coverage area, the condition, and the area for services provisioning and management are variable and seldom ideal.

The operation of an emergency-oriented wireless LAN would likely be based on using a mobile service headquarters that would provide the nucleus of the services and operate the transceiving equipment for the site. The mobile service unit would also establish the communication coupling with other remote or local services using traditional or wireless network connections. The local emergency site wireless LAN could be coupled to the remote LANs using wireless LAN bridging or wireless satellite communications.

Fast response

Speed of response and set of local communications services are paramount in an emergency services wireless LAN. Most of the units would have to be already set up and tested for their interoperability so they could be dispatched and unloaded at the site and put into immediate service. Any delays or special set up programming would limit, hinder, or even prevent the use of the wireless LAN systems. If they cannot help to orchestrate the emergency response services and improve the management process, then the systems would be nearly useless.

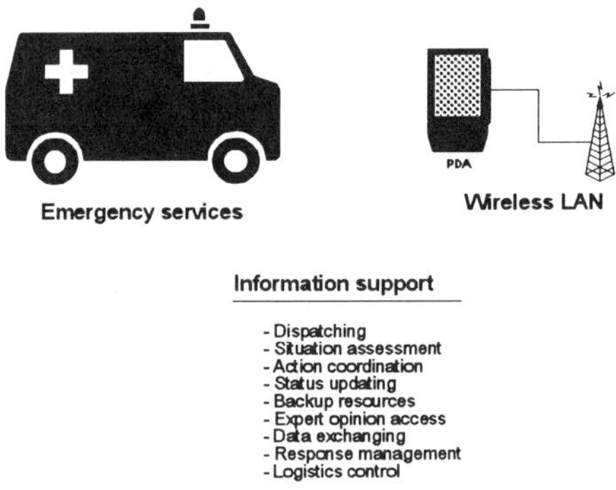

Figure 8.10: Emergency Response Wireless LAN System

Fast response can be aided by a complete deployment setup system for the wireless LAN. All units would be stored in a portable unit that could be dispatched to the emergency site with the advance contingents of people. Someone would be responsible for assessing the physical location of the emergency and implementing a dissemination plan to place, set up, and test the wireless units. Once operational, the units would be turned over to the appropriate emergency workers and the setup technician would move into operational and maintenance mode.

In addition to fast response for the establishment of the wireless LAN services, it is important that the system be highly reliable. The lives and actions of the emergency team may depend heavily on their ability to communicate to others on the network and/or obtain information and instructions from stored databases and action plans. If the wireless LAN cannot provide dependable service on a continuous basis, it should not be made a part of the emergency response systems package.

Global interconnectivity

The ability to go from the wireless LAN to a central communications hub which can in turn be coupled to existing networks (wireless and/or wired) is a key advantage of the wireless LAN.   It can be the conduit that allows any emergency worker to be in communication with anyone else in the contactable world. This could be extremely valuable in retrieving emergency condition information, remote expert advice, stored procedures, information on sites, terrains, past efforts, etc. All of this can be delivered from anywhere to the emergency site and then to the individual emergency worker in a timely and usable fashion.

The use of the facilities of the Internet and other emergency services networks would be the provisioning key to these services. The wireless LAN would provide the link from the individual emergency worker to a local intermediary server that would coordinate and manage the global interconnectivity.

Special and general services

The emergency support wireless LAN can provide a host of other services to the emergency workers and crews in addition to the access to emergency data and instructions. These services include:

- Task management

- Person location

- Operation control

- Logistics management

- Facilities control

- Provisioning for food and refreshment

- Scheduling of backup support

- Time tracking and triggering

- Messages and electronic mail

- Management reporting

- Media interfacing

- Bulletin board messages

The administrative support services may be inactive while the emergency is tended; as soon as the situation is under control the support services would be engaged and used to document and complete the summary of operations.

Radio frequency data collection systems

Radio frequency (RF) data collection systems are a logical candidate for the use of a wireless LAN as the communication coupling between the RF units and the data collection servers. These systems are usually based on handheld work units that can read and input data about task-level operations and status within an operational environment. The data from the units then need to move from the unit into a processing server, where they can be edited, checked, and added to the repository for further use and application.

Radio frequency systems are essentially a form of spread spectrum radio that is being used in a dedicated application format. For example, the handheld unit could be a portable bar code reader that can independently read the coded bars, interpret their data, and send a radio frequency message to a local server to process the data transaction. These are technically not true LAN systems because the format of the data is unique to the application and not standardized LAN packets.

Although this is a simple operation that uses a predominance of one-way traffic from the RF unit to the server system, these systems are heavily used and provide an important data link in more and more applications. They improve the data collection process, support improved mobility and performance, are reasonable in cost, and provide reliable service over a fairly long life.

Devices supported

The devices used in an RF data collection system are less capable than a typical personal computer client. The units are usually small handheld personal digital assistants or fixed function devices for data recording. Typical devices for an RF data collection system could include:

- Handheld inventory calculator devices

- Bar code scanners

- Voice entry boxes

- Small pen-pad computers with stylus inputs

The devices usually have a small display screen that can display the actions of the handheld unit and sometimes display instructions or data provided from the server level of the system.

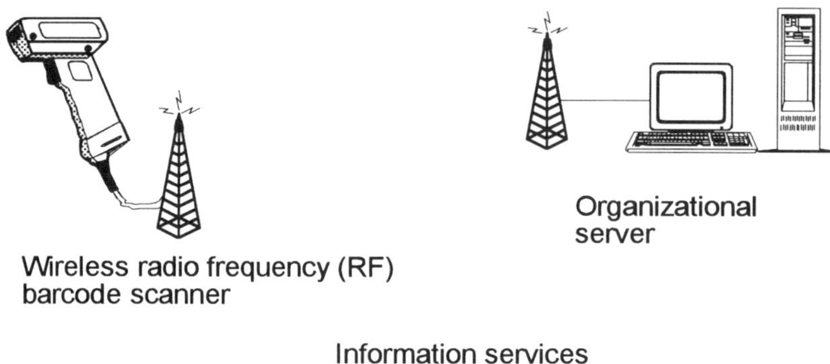

Figure 8.11: RF Data Collection Via Wireless LAN

Coverage requirements

The RF data collection systems usually have to cover a wide area of space—for example, a complete warehouse or a large retail store. The good news for coverage is that the space is usually fairly wide open, making for a good antenna and communications area. The bad news is that the space is large with some areas in which it may be difficult to provide reliable coverage. The use of repeaters and amplifiers and multiple hop services may be needed in these installations. Difficult space environments will require more careful design and setup as well as more equipment to properly send and receive the signals throughout the required operational spaces. The net result is that complicated and difficult environments could cost from two to three times as much as the cost of "clean" wireless environments. Although this may not preclude the use of wireless, the extra costs will make for a tougher and longer justification and payback period.

There may be some unique obstructions that will hinder the communications flows. For example, in a retail grocery store shelf inventory application when the data collector is working on floor-level shelves, the data collection unit may be out of the communications zone until the user moves it up into the higher spaces. Fortunately, the data units can operate independently on their batteries for the data entry efforts and need to be in the radio contact space only when they are transmitting and/or receiving data from the server.

Mobility support

The radio frequency data collection systems will need maximum mobility support. Units mounted on a mechanical arm of a forklift could go from the floor to the ceiling within their usage environment. The truck itself could easily move into spaces where it is out of contact with the wireless communications system. Because of the nature of the mobility and the environment, the RF data systems are much more difficult to restrict than other wireless LAN units. This places a heavy burden on the ability of the radio signals to be reliability maintained throughout the work space. One solution is to place more transceivers in a zone- type coverage arrangement so the RF units are always within easy contact with a transceiver. The zones would overlap one another and be of limited distance to assure good reception. This could double or triple the number of transceivers needed and add additional layers of intertransceiver communications and extensive control requirements.

Other solutions are to keep the mobile units more independent, allowing them to operate with stored information and to store their data in their battery-based memory until they are back within the wireless transmission coverage area. The systems would have to know when they enter the nontransmit zones and switch to operating in their independent mode. Another signal would have to indicate when it could return to the transmission process. Cellular systems management concepts are often used for these RF data collection systems.

Special services

The radio frequency data collection units are a specialized application of wireless LAN technology. They use the same frequency spaces as other systems, but their services are more specific and less user variable. This is both good and bad from a services point of view. The good news is that the system can be tuned to support specific uses and functions and security is not always a critical issue. The bad news is that these loads are heavy and demand high levels of accuracy.

Additional services for the handheld RF data collection systems could include the holding of significant amounts of data within the handheld unit and then transmitting data to the server at specified times or locations (for example, when all data are taken for a particular product category or the end of a row of shelves is reached).

Special environments

The use of wireless LANs in specialized environments can provide the value of quick setup, full services, fast mobility, and extended contacts. They can also operate in difficult environments that would not tolerate or accept wired systems. The following are examples of special environments and how wireless LANs could support them.

Hazardous situations

Hazardous situations such as a chemical spill, a fire, flood, warfare, etc. would all be amenable to the use of wireless LANs. The high mobility and the need for constant contact with other personnel and information resources would meet the capabilities of the wireless LAN. The client and transceiving units would possibly be susceptible to the contaminants of the hazardous situation, but they could be protected in the same manner as the humans and still provide their communications services.

Time-critical applications

Time-critical applications are those that require a short interval of time to complete. This could include the harvest of a crop or the support for a building demolition. In time-critical situations the wireless LAN would provide fast setup and operation of the connected information services. The ability to set up the service and then take it away upon completion would be a major asset.

Remote locations

Setting up a LAN in a remote environment such as a construction site or an emergency services location would be an ideal situation for a wireless LAN. Speed, portability, local services, and remote access would be the required services the wireless system would provide.

LAN-LAN couplings

Wireless LANs can also provide services directly to the internal parts of networks, rather than always to end-user situations. In these operations the wi reless LAN can cover longer distances, operate at higher speeds and perform in competition with many wide area network services (WAN). An example of this is the wireless LAN bridge which can be used to couple one LAN to another that is up to several miles away. Using line-of-sight paths and either radio beams or microwaves, these systems can transmit large volumes of data at high speeds with excellent reliability.

The LAN-LAN couplings are usually tuned and nonmobile. This means that once they are set up and validated they will maintain their alignment, can support higher intensity signal power, and simply use the air space as their working medium, rather than wire. Given that any outsider intervention with the signal would change the signal itself in easily monitored and measured levels, the overall security of this form of wireless LAN is very good. For those with some fear and reluctance to transmit sensitive data over open air space, encryption and decryption processes are availlable.

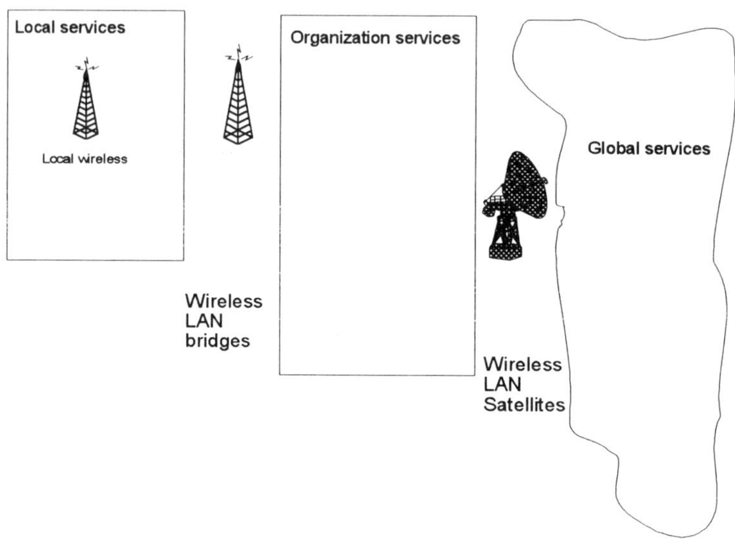

Figure 8.12: LAN-LAN Coupling

CHAPTER 9

Wireless LAN Standards and Protocols

Wireless LANs are still a part of the overall world of local area networks. This world has become heavily standardized by the work of the Institute of Electrical and Electronics Engineers (IEEE). The IEEE was requested by the International Standards Organization (ISO) to oversee the development and promulgation of standards for the LAN area during the late 1970s.

The IEEE 802 series

The basic 802 series establishes the specifications for the media access layers and the media access services. These define how the data will be organized on the LAN and how systems will access the LAN capabilities. As the wireless LANs are a part of the standardized LAN world, they will conform to the standards of the 802 services. The uniqueness of signals in the wireless environment will be supported by specific standards that will operate at level 1 of the OSI seven-layer standard. The overall organization of this standard is shown in Fig. 9.1.

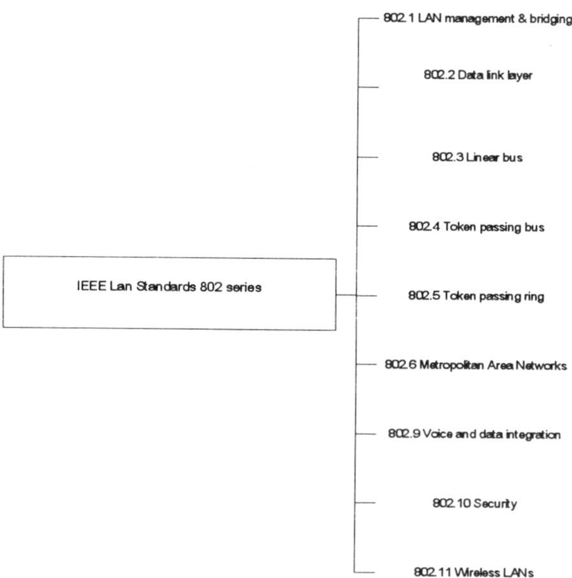

Figure 9.1: The IEEE 802 LAN Standards Series

Vendor standards

The final details of how a wireless LAN interoperates with its transceivers and the individual clients and servers will be left to the orchestration of the packaging vendor. Vendors will need to define the hardware, software and operational components of their wireless LAN product. Details of how the signals are engaged, transmitted, and received are outside the standard and can and will be set by the engineering of the producer. This will probably mean that a complete set of units that are vendor specific will be needed to institute and operate a wireless LAN. The interchangeability of transceivers and the ability of different wireless LANs to interoperate with one another are limited to nonexistent. The most likely level of common interchange would be to have a single server that is connected to the networks of two or more vendors.

This is likely to change as the official 802.11 standards become a reality and other vendors decide to license existing vendor approaches. Until it does, the wireless LAN world will be like the early days of facsimile, when a Xerox fax could send only to another XEROX fax. Today with standard CCITT Group fax standards, any fax machine can communicate with any other.

The 802.11 standards efforts

Within the IEEE 802 standards group a new section was setup to deal with the evolving issues of wireless LANs. This has been given the designation of 802.11. It has been meeting and collecting inputs since 1990. It has released some drafts of proposed standards. A first formal standard to be voted upon is expected to be released late in 1995. The concentration of this group is in the areas of:

- Defining the wireless interfaces

- Establishing standards for the signal levels of the different wireless LAN technologies

- Developing standards for the error management processes

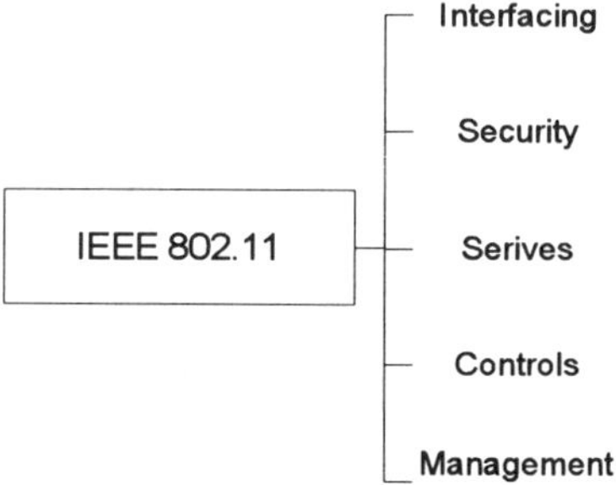

Figure 9.2: The 802.11 Committee

DFWMAC decisions

The 802.11 committee has internally adopted the Distributed Foundation Wireless Media Access Control (DFWMAC) as the foundation for the development of the overall wireless standards. This was taken from a proposal provided by AT&T Global Information Solutions/NCR, Symbol Technologies and Xircom, Inc. This proposal is to establish some standardization at the media access control (MAC) protocol level. DFWMAC deals only with radio frequency transmission in a spread-spectrum format. It also tends to establish a mechanism for separating vendor products in the wireless LAN spectrum rather than building interoperability specifications.

Minimum speeds

The original thinking of the working group on 802.11 was to select 1 Mb/s as the minimum speed level for wireless protocols. This was deemed adequate for basic file transfers and transaction processing. Further review and evolution of the technologies have changed this thinking to setting 2 Mb/s as the likely minimum bandwidth level. These are minimums and vendors are free to move to higher levels if possible.

Numerous vendors have felt free to develop their own speeds of information transfer over wireless LANs. This has produced a speed range that runs from 0.0192 up to 20 Mb/s. The minimum of the standard will probably establish the lower end of the wireless LAN spectrum.

TIA interfacing

The Telecommunications Industry Association (TIA) is a vendor-based organization that has also participated in standardizing some of the components of wireless communications. The TIA is involved in the establishment of time division multiple access (TDMA) technology. This is a technique for sending multiple digital signals over a radio frequency bandwidth. TIA is also involved in the definition of code division multiple access (CDMA). Much of the TIA involvement in wireless standards is at the invitation of the Cellular Telecommunications Industry Association (CTIA). Although TDMA and CDMA are primarily cellular standards, they can be used within the personal communications systems (PCS) world and on wireless LANs.

Signal coding alternatives

The signals that will travel over the wireless environment will require coding and integration together to create and maintain a service flow. Given that the signals can come from many sources and flow to different destinations, the overall signal coding and management will have to support some form of multiplexing. Multiplexing will in turn require rules to define how the signals will be packaged together and separated back into their original form. The wireless LAN world is considering two different forms of multiplexing, one using time slots and the other using coded packets.

TDMA

Time division multiplexing (TDMA) involves defining short time slots and then placing portions of different messages within unique time slots. Over a longer time, several messages are sent by interlacing orderly parts of the message into different time components. TDMA is like having several train tracks with boxcars coming together to form a single train that will travel over a single (bandwidth) track. TDMA transmitters will look for open time slots on a range of available frequencies. They will send the transmissions in segments using the available space. The TDMA receivers will need to listen to all used frequencies to find message parts that are addressed to their location. The receiver will have to assemble and organize the message and manage the quality through retransmission requests.

TDMA is a relatively well-defined process which uses the limited available frequency space to a high utilization level. It does place a high burden on the receiving station and can slow down if the network has a high error rate.

CDMA

Code division multiplexing (CDMA) was developed by the U.S. Department of Defense for secure voice communications. The technique breaks the inbound signals up into packets, each of which receives a unique code and is then transmitted over a wide band of frequencies. The receiver has to know what codes to look for and then retrieve and reassemble the messages by collecting them from across the transmitting frequency range. This approach provides greater security of the messages and supports a better demand and random use of the available bandwidth. CDMA also provides a higher utilization and greater level of traffic throughput than TDMA. The message slotting is faster and the error control is reported to be better. The problem with CDMA is that it is more complex than TDMA and many vendors have been having trouble producing a stable performance environment with CDMA.

Both TDMA and CDMA are very important to the cellular industry as they move to digital communications and to finding ways to handle higher traffic within their limited bandwidth. As the cellular standards evolve and the technologies reach hardware-embedded stability, they will move over into the wireless LAN world.

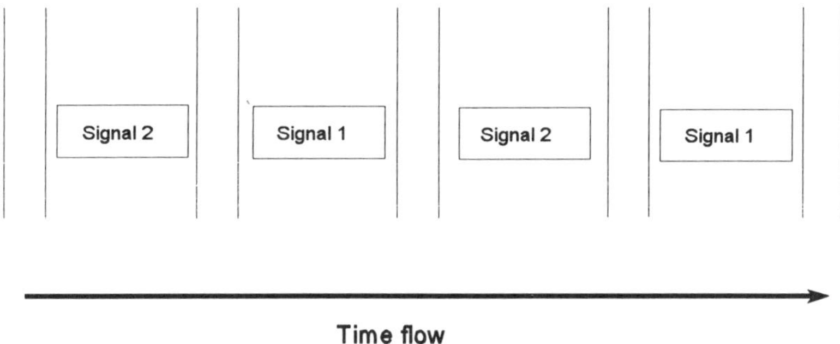

Figure 9.3: TDMA Multiplexing

International considerations

On the international front the standards efforts for wireless LANs are placed in the hands of the established communications agencies. These are heavily populated by members of the European Post Telephone and Telegraph organizations. This gives a heavier governmental character to the European standardization efforts than is found in the United States.

European Telecommunications Standards Institute (ETSI)

The ETSI has dominated most of the European efforts to standardize communications products. This is especially true since the formation of the amalgamated concept of the European Economic Community (EEC). The ETSI was the body instrumental in establishing the Digital European Cordless Telecommunications (DECT) standard and believes that any use of the cordless spectrum comes under their responsibility.

Even with the EEC, the actual control of standards and air wave allocations are still defined by the individual governments. Their sovereignty still reigns as the final word in what can and cannot be done within various countries. For example some of the spread spectrum space used in some countries of Europe is *not* available in others because it has been allocated to other uses. It is thus important to check on each country before committing to a wholesale implementation of wireless LANs.

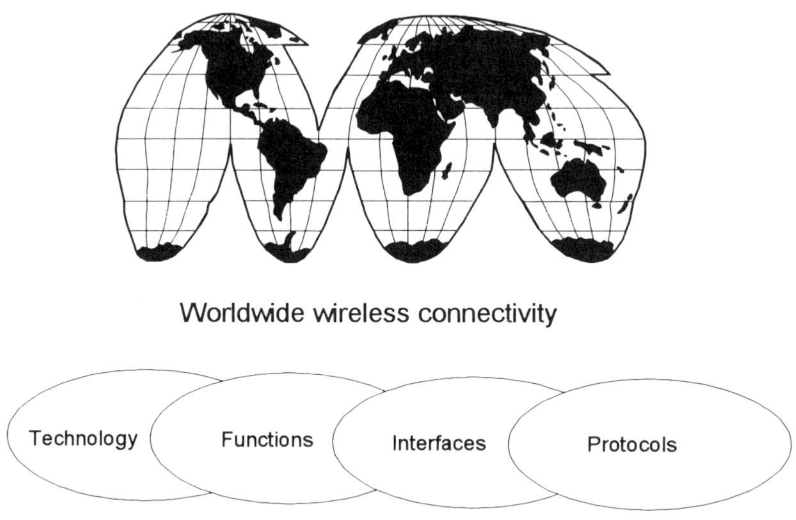

Figure 9.4: International Wireless Standardization

The good news is that wireless LANs are local within a building site or they cover a regional bridging space. Few wireless LANs will span international borders. If they do carry traffic that must cross international borders, the wireless LAN will probably hand the traffic off to the WAN system to perform the transfer using already established networking connections and control practices.

HIPERLAN

The High Performance European Radio LAN (HIPERLAN) is an ETSI project to develop a common set of frequencies that will be available for low-power spread-spectrum communications across Europe. They may or may not agree with the available bandwidth in the U.S./North American community. HIPERLAN efforts should be monitored to determine the availability of bandwidths within the European setting. Today one has to conform to the country-specific standards for wireless LANs. However, it would be an expensive proposition to do something today and then find the HIPERLAN defines different frequencies that require users to change what they just put in. Fortunately, governments do not agree or legislate new technology rules in a speedy manner. This means that even if you choose the wrong technology for your European wireless LAN you would have several years to obtain results from your investment before having to replace it with the new mandated technology.

Interoperability standards

The wireless LAN will usually operate as a subsystem of other LANs. This means the interconnections and the interoperability between communications systems and different LAN technologies will be a critical area for standardization. This will be on the longer term agenda of the IEEE 802.11 committee, meaning that it will be largely left to the vendors to define and develop the mechanics and procedures for interfacing to and from other worlds to the wireless LAN environment.

LAN-to-cellular futures

The wireless in-building LAN may serve as a logical extension of the remote mobile person. Individuals would use a cellular connection to support their communications during their geographic travels and then would automatically connect to the in-building wireless LAN world once they were in the building environment. Connections from cellular to wireless LAN would be accomplished by using some form of data packaging scheme on the cellular system. Current analog modems for cellular services are notoriously error prone and costly. The short-term hope is for a new protocol known as Cellular Digital Packet Data (CDPD) which still uses analog transmission but using packets that are in ditigitized form with better error management. The protocol is a form of old X.25 and operates at 19.2 kb/s. Widespread implementation and availability of CDPD should be completed by the end of 1996.

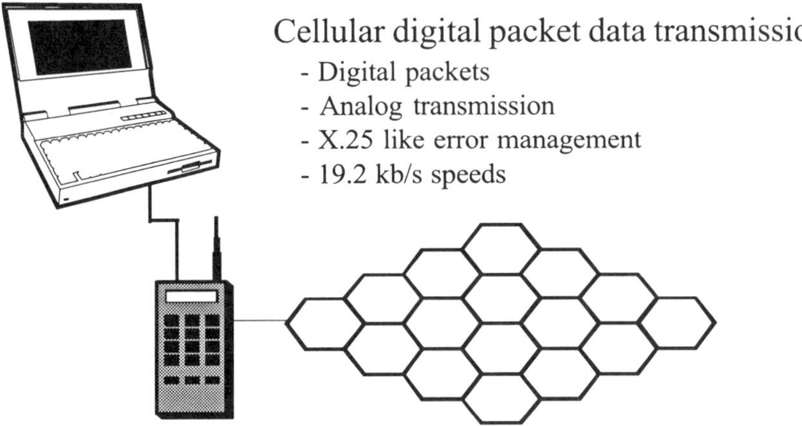

Figure 9.5: Wireless CDPD Transmission

The connection of the cellular world to the wireless world will be through an intermediate server that would act as the transceiver between the two different wireless technologies. The connections will operate in the same way as they would when the cellular services access a wired LAN environment.

Users of a cellular to wireless LAN connection would be organizations that had an internal and an external mobile workforce. This could include trucks with cellular data systems that couple to a freight warehouse or distribution center. Field service technicians could use cellular data to communicate with a mobile technical staff of backup experts or a parts ordering center that uses wireless LANs. The largest market for these services could be the mobile sales forces.

The wireless LAN to cellular connection can deliver transactions at the time of occurrence, greatly reducing time delays. They can provide near-instantaneous access to stored databases to the field personnel, who can then make site decisions on the situation and transaction. Although the databases (in the servers) will probably be fixed within a wired world, the wireless LAN and cellular connection makes the data available to both a local building environment and the remote mobile field forces.

Wireless to wired

The connection of wireless to the wired LAN world is the most common linkage. This permits the wireless system(s) to supply the mobile workforce with LAN services and then allows them to interface with the established (wired) services of the organization. This connection is maintained via a shared server that is on both the wired and the wireless networks.

In many instances the connection between computers is from wireless client to wired server to a wired control computer to another wireless client to a wired server back to the wired control computer and finally to the originating wireless client. This may seem a convoluted approach, but the wired connections are still cheaper than wireless and if the local clients are close together, the use of a small daisy-chained wiring setup is cost and performance effective. This limits the data traffic to that which must leave the local work group and move over the larger facility space to the server world. This also avoids the problems of excessive contention for the wireless bandwidth and any overlap of signals that might exist when several users in a small area all try to use the same wireless network.

The key to the decision about where and when to use wireless versus wired can be determined by the Negroponte switch (named after the originator Nicklaus Negroponte of the MIT Media Lab). The Negroponte switch says that if the users move they should be coupled via a wireless network and if the users are fixed (house, database, file cabinet, etc.), then the unit should be wired. Although simplistic in nature (KISS theory), this decision advice often works to support the technical and operational effectiveness of the situation.

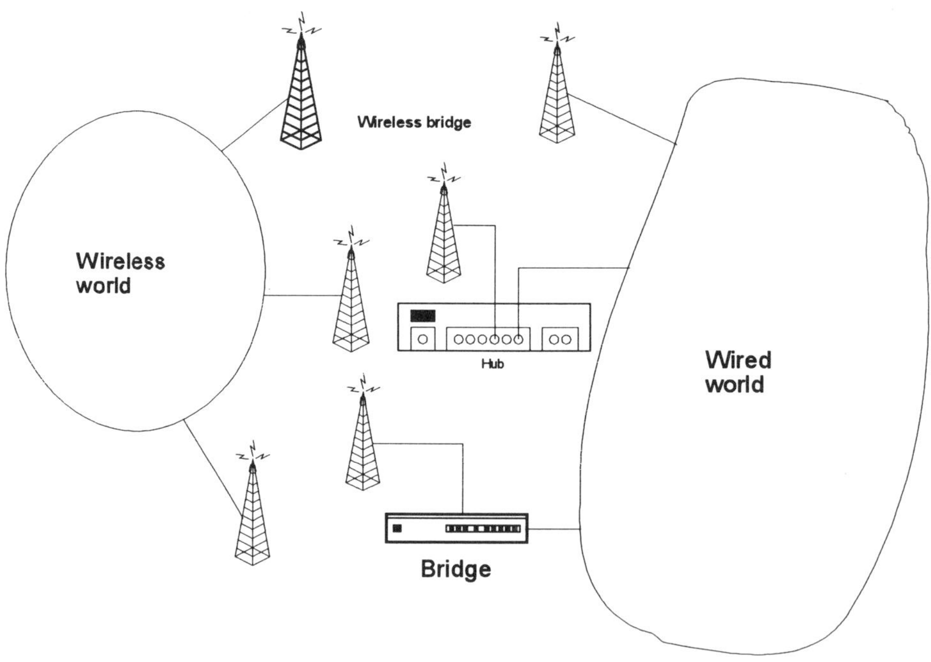

Figure 9.6: Wireless-to-Wired LAN Connections

Software couplings

Software couplings for the wireless to the wired LAN can occur at several levels. The most logical layer is at the network operating system (NOS) level. It could also occur at the application level. For example, to expedite the transmission of data in the pit broker applications, the communications coupling are written into the applications and there is no network operating system. This saves time and costs but makes the application rather proprietary.

Software couplings can also occur through the use of middleware in a client/server type of relationship. In these cases the middleware receives the wireless communication and determine where and how to move it to the appropriate response server station.

Packet management

Packet management is at the lowest data level within the wireless LAN world. The packets are in the format of the established framing convention being used (Ethernet, token ring, ARCnet, etc.) As the packets are generated by the computer station, they are identified and managed as individual units through the wireless LAN world. While in the wireless LAN the packets retain their standard format; however, the wireless LAN system could add other components or data elements that are not standard as the systems are vendor specific and not interchangeable.

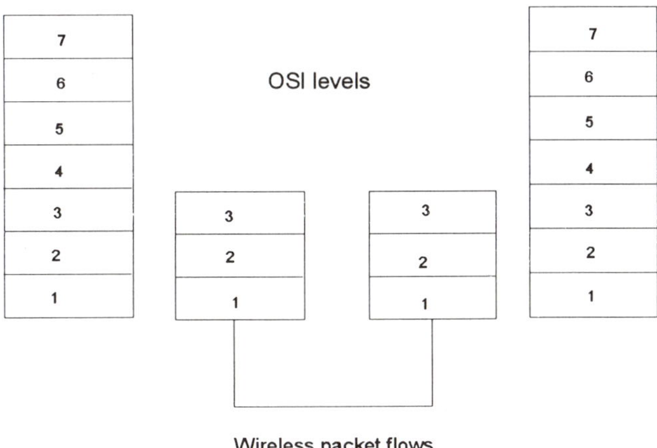

Figure 9.7: Wireless LAN Packet Flows

Wireless standards evolution

As wireless LANs establish their presence in the application world, their standards will become more important. The evolution will be to develop the wireless LAN as a part of the overall LAN world. It should not become a separate component that is pasted into the rest of the networking world. The wireless LAN is another choice of mechanism for moving data between users and organizational components. It is a good solution for certain types of systems and situations. With the standards integrated across both wireless and wired worlds, users can expect to receive transparent services to their LAN requests and not be bothered whether the services came from a wireless, wired, or mixed world.

CHAPTER 10

Wireless LAN Decisions

To use wireless or not to use wireless, that is the question. Rather it is one of the decisions that have to be made. If the answer is no, then the rest of the decisions are unnecessary and can be avoided. If, however, the answer is yes to the use of a wireless network, then there are many issues to be resolved. This chapter defines the range of decisions to be made for a wireless LAN installation and operation and concentrates on defining the factors that will maximize the success of such an installation.

Usage

Usage is the ultimate decision factor for defining the specific choices of technology and orchestration of a wireless LAN. The kind and type of application and the data to be transferred back and forth on a wireless LAN will be the keys to the technology selection. Short bursty transmissions will suggest radio spectrum, long file transfers will suggest infrared, and long-distance, high-volume transfers will suggest the use of microwave services. Mixed form usages will be more difficult to specify and the dominant one will likely be the decision influencer.

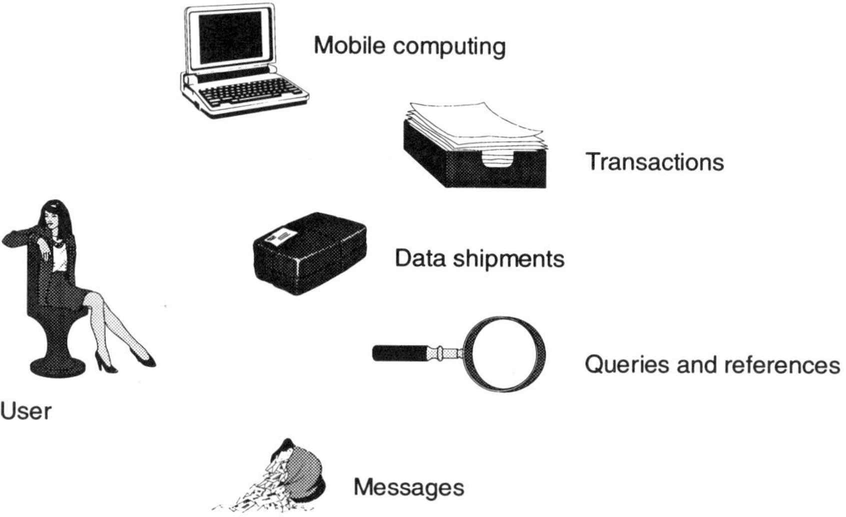

Figure 10.1: Usage Considerations

In situations where usage varies over time or with changing applications, the decision logic will be more difficult. The issues of volume, type, and sizes of transfers will then become the determinators. The greater the volume of transfer, the more the suggestion will be to use tuned or fixed wireless like infrared. These technologies can support higher speeds and more sustained throughput than the spread-spectrum radio beams.

Locations to cover

The locations to be covered in the wireless LAN environment will also play a key role in the decision process. The distances between users and the coupler nodes and the physiology of the space will be important factors in the design. Large open areas are preferred to tight, filled offices or high racks of warehouse space filled with heavy goods.

The wireless signals travel better with less distortion and interference through open air spaces and will deteriorate or disappear if there are too many items in the way of a clear signal path. If the space to be covered with wireless signals is too large or the signals must pass through too many walls, doors, arches, etc., the wireless approach may not be feasible.

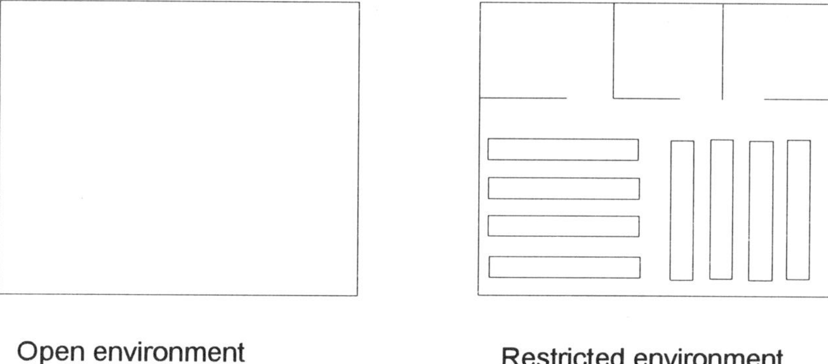

Figure 10.2: Impact of Location Variations

Facilities

The space and physical facilities that will house the wireless LAN are a critical factor in the specification, design, and implementation of a wireless LAN. As the wireless LAN signals are affected by the surroundings and the area in which they must be maintained, the facility is key to the success or failure of the wireless technology. The age, makeup, condition, use, structure, services, and other parts of the physical facilities must be evaluated prior to determining the usability of wireless LAN technology. If this cannot be done due to lack of information or time, then the only other reasonable option is to set up and run trials of the technology in the areas in which it will be used. As wireless is a quick, minimum setup technology, this may be the most expeditious and proof positive way to go in many instances.

Where facility information is available and accurate, the items shown in Figure 10.3 should be evaluated for a determination of the likely usability of the wireless concept.

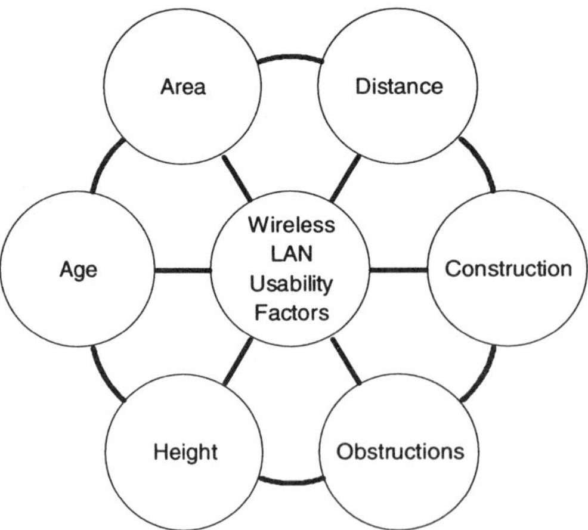

Figure 10.3: Facility Evaluation Factors

Condition

The condition of the space and facilities is a major criterion. If the area is piled up with goods and equipment, disorganized in general, subject to major changes in character, etc., then the wireless LAN technology will be easily subverted and made useless by the regular changes that occur in the facilities. Messy conditions are also a hindrance to good open-air signal management and could disrupt the communications.

The best conditions for wireless LANs are clean, orderly, predictable conditions with minimal chance of out-of-the-ordinary items entering the communications space used by the wireless LAN. Open offices, low-level warehouses, retail stores, person-oriented laboratories, and similar facilities make for the settings and conditions in which wireless LAN technologies can operate successfully.

Space utilization

The usage of the space and facility is also a key factor for wireless LANs. Office use is better suited than some heavy industrial usage. A library is better than a warehouse, and a research laboratory is usually better suited to wireless technology than a machine shop. Although these are generalities, which can sometimes be overcome with good design and layout of the wireless LANs, usage patterns that are more conducive to the technology will help assure the success of applications.

The usage of the facility will also tend to influence the use of the wireless technology. A large open roaming space, such as in a manufacturing plant, will be harder to cover and operate in than a small office department. Usage of the facility will also establish the location of the wireless transceivers and the availability of open channels of signal movement.

Age

Age should not be a factor in most systems. However, in wireless technologies, the age of the building facility space is critical to the ability of technologies to move their signals clearly and reliably. Older (possibly historic) buildings tend to have thick walls and heavy superstructures, plus layers of modifications. All of these items tend to restrict or impede the flow of wireless signals, making it more difficult to setup and successfully operate the wireless systems.

Newer building are strong but use lighter weight structures and materials. The interior walls are thin partitions that permit passage of most wireless signals. There is also more open space in newer buildings, which rely on external shells for support rather than massive structures of heavy interior walls.

Prior modifications

The history of modifications to a building facility is also important for wireless LAN planning. If there are many layers of modifications to the physical space there is a greater likelihood of hidden conditions that could restrict or interfere with the wireless LAN signals. Older buildings often have heavy steel framing which has been covered over with plaster or paneling. There might be covered-up doorways with heavy hidden girders, hidden electrical conduit piping, old gas piping, and many other hidden elements that might adversely effect the wireless signals.

The best approach is to obtain some insight into the history of the building facility and the types of changes and modifications that it has undergone since its origin. This information may be difficult to come by and may be vested in the minds of some of the building managers who had involvement or interest in the history of the facility. Reasonable attempts to secure the information on the prior modifications should be made, but in many instances the results will be limited.

If there is any doubt as to the appropriateness of past modifications and their influence on the use of wireless LAN technology, it is advisable to do a physical inspection of the underlying elements of the building and/or run tests on the use of the wireless technology.

Physiology

The physiology of the building or facility space where the wireless LAN is proposed to operate should be studied and evaluated. There are usually floor plans and building layout diagrams that can be obtained from the facility managers. These will provide the basic physiology of the area. The review of the physiology should cover the services distribution of heat, lighting, and ventilation and the layout of fixed partitions that could interfere with or limit the natural travel of the wireless LAN signals.

The physiology of the proposed wireless LAN and the spaces it must cover should be diagramed onto the facility space plan. This will allow for detailed inspection of any limitations or blockages that might be a hindrance to the movement of the wireless signals between the users and the required systems connections.

Spatial configurations

The spatial considerations for wireless LANs involve mapping out the space to be covered by the wireless LAN signals. The spatial configurations should consider the dispersion of the signals from the transmitting antennas and the shape of the wave moving through the surrounding air medium.

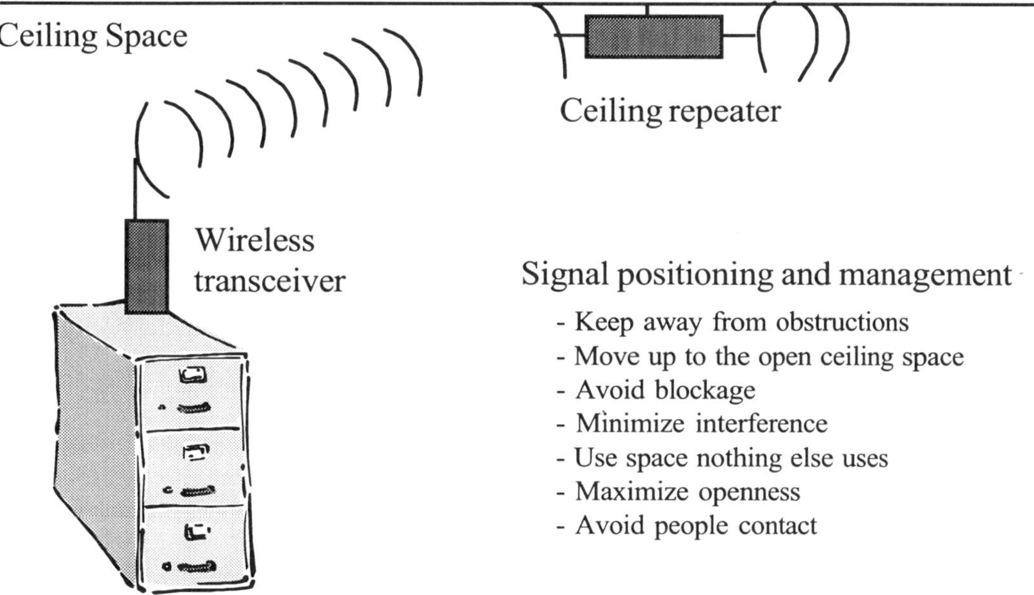

Figure 10.4: Spatial Coverage Options

The spatial area of coverage will be a fan-out configuration from the transmitting antennas. It can be affected by things that are located in the path of the dispersion wave. For example, metal filing cabinets, metal ceiling hangings, steel beams in walls, and even wall pictures can induce variations in or distortion of the patterns of signal dispersion.

The actual configuration of the space to be used for the wireless LAN should be inspected, mapped, and, if needed, cleansed of possible problem components. Open spaces, such as a trading floor of a stock exchange, provide better spatial signal management areas than most office areas.

Basic Service Areas

A basic service area (BSA) is the localized loop where a client computer reaches the coupling with the wireless LAN. This area is very distance limited and represents the first layer of connectivity to the wireless LAN world. The basic service area may be wired for cost and control efficiency; but at the end of the basic service area, the user's signals will be converted to the wireless system.

Extended Service Set

The extended service set (ESS) is the region where the signals travel around the facility space using the wireless format and technology. This would include multiple spatial areas with individual wireless coverage where the signals can hop from one cell to another via wireless connections. The total coverage area would be the extended service set and the individual cell would be the basic coverage area.

Wire versus wireless

The wire versus wireless decision is usually decided by the requirements for total mobility. The more the mobility factor exists, the greater the support for the wireless systems and services. As long as the physical space and geography will support a wireless configuration, it is usually the way to go.

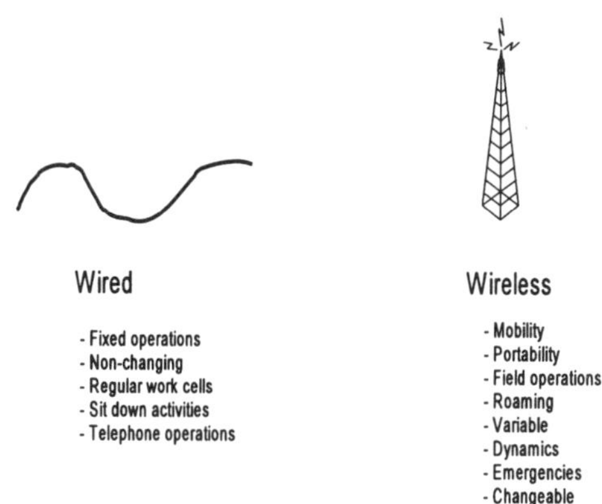

Figure 10.5: Wire Versus Wireless Factors

The wired LAN is not likely to give much way to the wireless technology. In most installations the wired LAN provides faster and more reliable transmission of data between the units. There would be little justification for removing a wired LAN and replacing it with a wireless system.

The relationship between wireless and wired LANs is one of needed partnership. The wireless LAN allows users to have mobility and connectivity that are not possible with the wired systems. The wired system provides the backbone architecture that connects more enterprise resources and provides a major level of service to and from the wireless users.

Points of coupling

The points of coupling represent the locations at which the wireless LAN can be connected to the wired world. These points can be existing departmental servers, special communications couplings, or added servers dedicated to making the connection between the LAN types.

Points of coupling represent the locations that the wireless communications must reach in order to attach themselves to the rest of the real world. The transfer point between the wired and the wireless technology will be:

- A traffic load point

- A conversion location

- In need of management

- Error prone

- Subject to delays and saturation

Points of coupling can be multiple on a wireless network, just as a wired network might have several bridges. Multiple points of coupling provide for alternate routings, backup in the event of failure, and more options for smooth traffic management.

Message conversions

The traffic on the wireless LAN will be packaged into some type of message packet. This may be standard Ethernet or a nonstandard vendor-defined message format. In most cases the messages will need to be converted in format from the wireless LAN to the wired environment.

Even in coupling directly to a wireless server, the transmission message has to be assembled from the packets and made whole for the information processing application. There will be some level of conversion of messaging going on at the point of presence and the coupling to servers or wired environments.

Performance

The performance of a wireless LAN depends on many factors. Although the rated speeds are slower than those of the wired LAN technologies, there are some trade-offs in favor of the wireless approach. One of these is that wired systems are contention based or act as a party line with only one (baseband) channel available to be shared between all of the users. In the wireless LAN concept the basic channel speed is lower, but there are several channels that can be allocated to server- requesting users. In addition, with the frequency hopping process, as the message travels from hop to hop, the prior hop frequencies are available for reuse by the next user.

Performance issues

Figure 10.6: Performance Issues

The overall performance of wireless LAN systems will need to be evaluated in the context of the exact layouts, configurations, channels, and other parameters, before determining the performance of the total system. The possibility of interference, errors, and other transmission disruptions need to be considered to determine the overall effectiveness of the wireless LAN application.

Throughput

Throughput is the net amount of "good" data that can be transmitted and received at the application level. The calculation of throughput also needs to consider the parallel use and loading of the network and not determine the net on only one application. This is especially true in wireless LANs, where multiple communications activities supporting several applications can be under way at the same time.

It may not work out as exactly balancing a 2 Mb/s wireless network delivering the results of a wired 10 Mb/s network, but the results can be better than the apparent 5:1 reduction in overall throughput. In some instances in which the wired traffic is "hogged" by a single user, the more open communications flow of the hopping wireless system may give the users more apparent service. Again, this will depend on specifics of the setup and the usage of the networks.

From a planning point of view, these uncertainties in the actual estimation of throughput in a wireless LAN world will suggest the use of simulation modeling and load analysis to define the expected throughput and service levels from a wireless LAN installation. One cannot overemphasize that as all LANs are user demand oriented, the actual requests from these users over time will determine the performance, results, and throughput of the system more than the actual technology components.

In a similar consideration, if the system (either wired or wireless) is poorly installed and subject to interference and errors, then the net "good" throughput will be lowered. This is more important in the wireless world, but it is not a nonissue in wired installations.

Speeds

Speeds are important in all of computing. Networks unfortunately (except for the human actions) turn out to be the slowest link in the equation of end-to-end information delivery. Therefore the raw speed of moving bits of data along the network is an important design consideration. The wireless LAN world has seen a constant increase in the speeds that can be carried; however, the technology has more severe limits than found in other media such as copper wire and fiber optics.

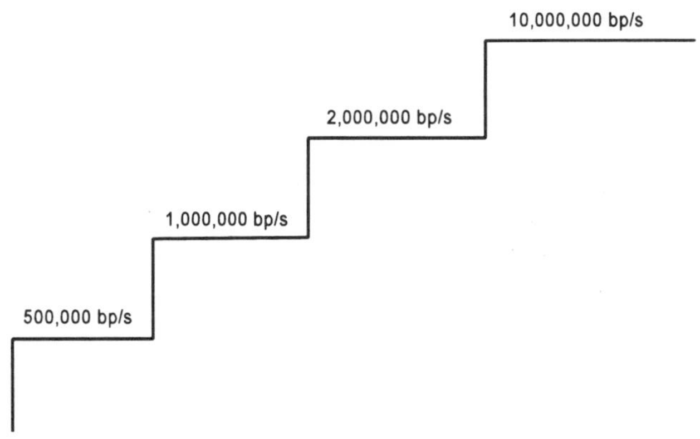

Figure 10.7: Wireless LAN Speed Comparisons

When wireless LANs were first introduced, their speeds of 115,000 bits per second (b/s) were only slightly faster than those of dial-up telephone communications. They essentially matched the speed of a serial computer port and were thus used to support printer and simple file displacement operations. As the speeds increased to 200,000 b/s and then to 500,000 b/s the wireless LANs could take on more traffic, but it was still mostly a supporting service type of movement such as file swapping, backup and printer traffic. Transaction and user waiting for delivery traffic was not the key load factor.

The next level of speed raised the performance to over 1,000,000 b/s and began to be of the same order of magnitude as that of other LAN networks. In fact, this put them in the ballpark of early wired LANs such as ARCnet and AT&T StarLan which began at 1 Mb/s. Unfortunately, the wired LANs quickly moved to 10 and 16 Mb/s levels and the wireless layer had difficulty in moving to these plateaus.

This has been partly rectified by the move to define the 802.11 standard as 2 Mb/s. The infrared group has been shooting for 16 Mb/s for their new level of wireless intercommunication. These will establish new performance and speed plateaus; however, they may represent the near limits of these technologies given the restrictions of existing bands in the communications spectrum. These twofold and 16-fold improvements will be welcome, but the wired LAN world is moving to 100 Mb/s and beyond.

Speed will always be an issue for the wireless world. However, remember that the wireless world a sublayer of the wired world. It coexists with and supports, and does not compete with or replace the higher speed world. These lower speeds can also be segmented to support fewer users at the local level. This would reduce the contention for the limited speeds and provide each user with a higher net throughput level.

Speed is technically important in networking. However, the net amount of service to the users is more important. With good design and limitation of competition for the limiting speed, a slow network can provide good services and satisfy user needs. This is a better measure of success than the engineering speed level.

It should also be remembered that the wireless LAN is being measured against wired LANs. If one were to consider that some organizations move massive amounts of data over 9600 b/s, 14.4 Kb/s, 28.8 Kb/s and 56 Kb/s communications lines and yearn for a T1 coupling at 1.544 Mb/s, then a 1 to 2 Mb/s wireless LAN is a very competitive and attractive alternative.

Errors

Nobody wants errors in anything, especially their data. As the wireless LAN has more points of susceptibility to errors, there must be more consideration and steps taken to keep these errors under control. There is good and bad news in the error management of the wireless LAN environment. The bad news is that there have to be more tests and validation and error correction services applied to the movement of data over this environment. The good news is that there is time in the equation to do these steps without reducing the net speeds and service throughput of the system.

Error management in wireless LANs is similar to than in other network services. Cyclic redundancy codes (CRCs) are added to data packets and then checked along the transmission route for correctness. If errors are found, the first step is to re-compute the correct data (CRC); if this is not possible, the packet in error is located at the source and retransmitted.

Signal Errors

Figure 10.8: Wireless LAN Errors

The concept of error management is that the system must assure the user that transmitted and received data are highly accurate and contain minimal errors. Zero error is not possible in any network system, so further checking at the data levels through audit and control techniques is always recommended, based on the accuracy needs of the data.

Connections

The connection to a wireless LAN is done mostly through direct request and contention access. About 85% of the wireless LANs follow the Ethernet contention and thus provide carrier sense multiple access services. The collision detection algorithm process varies with the different vendors products.

The wireless connection is instituted through the transceiver module at a coupling location. This module may be dedicated to one user or it may be shared by a local work group network (usually wired.) When the transceiver receives a request for LAN services, it will establish the signal packaging formats and integrate the user data into a transmission process. It will stay tuned to the receiving station to determine that the transmitted data are received correctly and without detectable errors.

Disconnects

The disconnection from the wireless LAN is done by completion of the transmission sequence. When the wireless transceivers have no requests to transmit, they will send some background packets to validate the availability of the circuits. However, when the receiver indicates that it has received the data from the transmitter, the system disengages the wireless network.

There is a chance for a dropout in the transmission over the wireless medium. This could cause a false disconnect and usually an incomplete packet transmission. In these conditions the transmitter would have to reestablish the communications link and retransmit the data packets.

Network management

The decision making for wireless LANs should also consider how the management of the wireless LAN technology will be integrated with other network management efforts. There will be some areas where the wireless LAN will need some unique management efforts, but the overall control of uses and applications should be closely integrated with other systems and network management procedures.

This section will identify some of the unique network management issues for wireless LANs. Chapter 13 will go into more depth on general network management practices and procedures for the wireless world.

Equipment failures

Wireless equipment is very reliable, as is typical of today's modern electronic circuits. However, it can fail at times and must be monitored in unique ways to assure that the overall system and service are maintained.

Wireless LAN equipment failures consist of:

- Power failures

- Bad packet generation

- Broadcast storms

- Transmitter tuning

- Antenna failure

These failures will need to be monitored at either the physical or data level of operation. When errors or failures are detected the unit will have to be identified, isolated, and replaced. Repair and replacement are usually done at a maintenance depot and not at field sites.

Network failures

Wireless network failures usually occur when the transmitter and the receiver cannot communicate with one another. The lack of a physical intervening media (except for the air) prevents the failure of the medium. But the generators and acceptors of the contents of the medium can be out of tune or off frequency or underpowered so that they cannot complete the wireless LAN communications.

If the hardware is operational and the messages are not getting through, the problem is usually with the transceivers on the network. The only other network-level failure would be some type of physical interference with the signal path, such as shadowing by people or a change in the physical space.

Tuning

Tuning is critical in all of the wireless LAN environments. If the signals cannot reach between the transmitters and the receivers, the system is in trouble. Tuning failure will usually bring the wireless LAN to a complete stop.

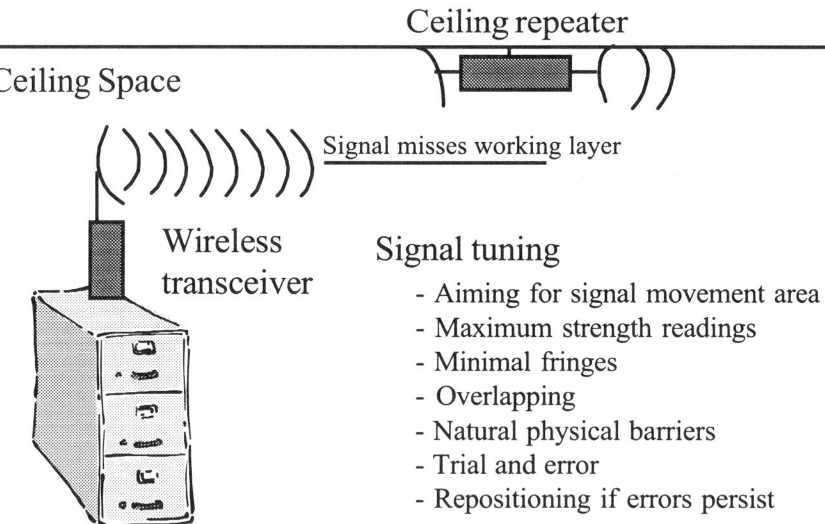

Figure 10.9: Wireless LAN Tuning

The maintenance effort for tuning of the wireless LAN involves regularly testing the transmitter and receiver linkages and the frequencies of the transmission and ascertaining that they are within specifications. When they begin to vary the maintenance effort will need to retune them and validate the overall operation.

Fault detection

The fault detection in a wireless LAN should be conducted as part of the equipment operation or it should be integrated into the overall network management center for the enterprise. There is little need for maintaining a separate network fault detection process for the wireless world.

When the fault detection is built into the equipment, the reporting and addressing of the fault should be part of the overall network management schema. Where there are tests to be performed to determine the health of the wireless LAN, these can be performed on a remote basis by the network management center.

This is not to say that there will not have to be some training or trained individuals who understand the concepts and the operation of the wireless LAN equipment, however, they can be members of the network management organization.

History tracking and reporting

It will be useful to identify the kinds, types, and frequency of errors and faults in the wireless LAN environment. This is no different from wanting to know the same information for any segment of any network. By keeping such statistics and data, the management can address problems on a proactive basis and avoid major failures of unacceptable reliability from the systems.

Wireless LAN management efforts

Remembering the advice that the wireless LAN management should be a part of the overall network management process and not a separate and independent unit, this section will define the management efforts within the context of the network management process.

Monitoring equipment

Some special equipment will be needed to monitor various types of wireless LANs. This should be added to the kit of the network technicians just as cable testers, protocol analyzers, and other equipment have been for other networks. The unique test equipment needed for wireless LANs includes:

Transmission testers

Devices that can read the communications values from the air medium and determine whether they are correct.

Protocol analyzers

Devices that can dissect the air wave communications and determine that the data packaging on the wireless LAN is correct.

Management Software

The software applications that will run on a wireless LAN and/or the network operating system version will likely possess additional testing and maintenance facilities to allow the network management group to evaluate the data- level services of the LAN. The vendor- supplied driver software often has capabilities to do the following:

- Self-test the units

- Trap errors and report on them on request

- Maintain statistics on the loads and performance

- Provide audit trails for defined segments of the network

As management and operating systems software advances to the next generation, the wireless LAN operations should be amenable to the plug and play concepts and to imbedded network operations support. Indeed, new systems such as Microsoft Windows 95 will be providing support for wireless network services in both their connectivity, plug and play, and network operations components. Support for remote service requests and user roaming will also be provided.

Use tracking and reporting

The usage of the wireless LAN can be variable depending on the application and the users. It is useful to develop some true sense of how much traffic and what type of traffic is moving over the wireless LANs. This can be done by maintaining summary usage reports of volumes to and from various stations and the types of traffic handled.

Most network operating systems will provide detailed audit logging capabilities that will track and report on each transmission. This is useful only when chasing down errors at the message level. The building of summary statistics by type is more than sufficient for the usage analysis process.

CHAPTER 11

Wireless-to-Wire Connections

Very few wireless LANs operate only in the wireless mode. The usual configuration is to have the wireless LAN operating as the local connectivity layer for the mobile users and then couple the system to a wired LAN via a common server node. This provides the best of both the wireless and wired worlds. Users at both ends of the spectrum can have access to shared resources and each other.

Wireless-to-wire LAN connections are relatively easy to set up and maintain. A selected server is configured to be a part of both the wireless and wired LAN environments. This server functions as the intermediary for both environments.

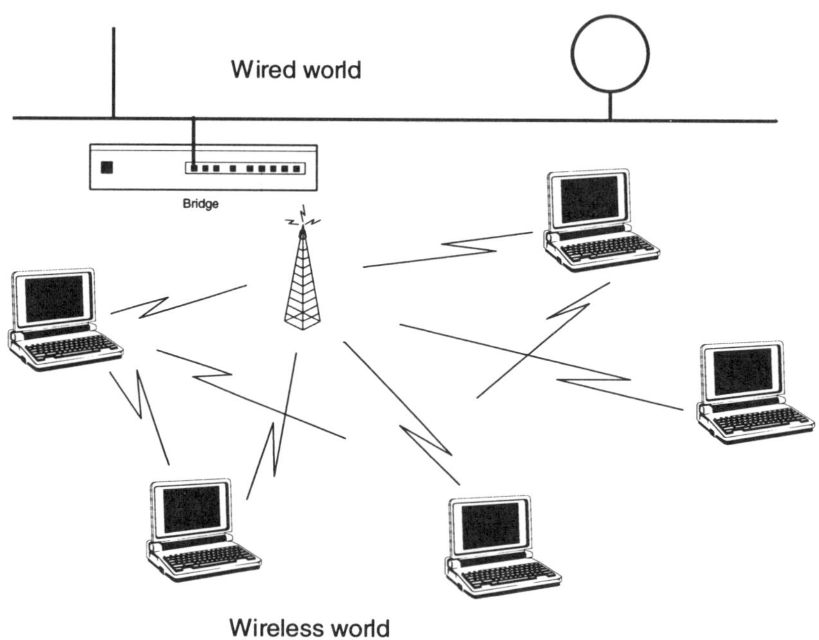

Figure 11.1: Typical Wireless-to-Wire Linkage

Remote access

The provisioning of remote access to wired networks via a wireless link involves a
tightly managed relationship. The provisioning of security controls and careful
screening of the data flows is needed to maintain the integrity of the environment.

Remote access will provide flexible access to a broad range of services, however the remote access connection will probably be slower and more error prone than local couplings.

Resource manager

The resource manager will have to identify the request of a remote wireless user and determine where and how to provide answers to the request for services. The resource manager will have to know:

- Authorized users

- Location of resources

- Access codes and directions

- Service possibilities

- Alternate routings

- Validation and security permits

The resource manager will probably be vested at the wireless-to-wired interface point or in a shared server that is somewhere in the wired world.

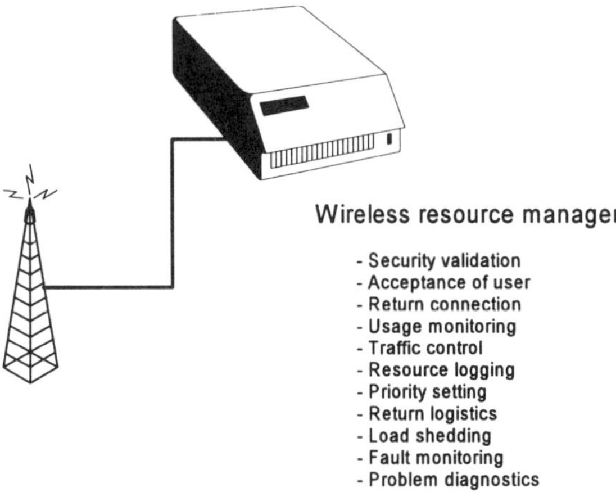

Figure 11.2: Resource Manager Support

Bridging and routing

Bridges will allow traffic to pass from one network to another. The bridge is nondiscriminating and will pass all of the traffic presented. A bridge takes traffic and moves it to the other side of the bridge onto the connected network. In wireless LANs, the bridge may be connecting the wireless world to the wired environment. Bridges could also be used to couple one wireless work group to another.

Routers are more intelligent than bridges, in that they contain tables of legitimate addresses and network locations. When presented with a message, the router will examine the address and determine that the addressee is a known entity in the network. If the address is unknown, the router will eliminate the message as erroneous. If the message is bound for a remote network, the address would be a known connector port within the router's knowledge space.

When a wireless network hands a message to a router, the unit could determine the best way to proceed with the message's movement in the network, as the router also knows the links and conditions within the network. Routing would seek to send the message by the most efficient and least congested connections.

Bridges and routers will be natural points of interconnectivity between wired and wireless networks. These devices' natural role in interconnecting networks will work in both wireless and wired media. The use of standard protocols (Ethernet and token ring) within the wireless world will allow the bridges and routers to function as a wireless coupled device with minimal change.

Central servers

Within the wired world there is a tendency to build and maintain some central repositories and servers. Examples include data warehouses and electronic image filing cabinets (optical jukeboxes). These central servers can be coupled to the wireless world through the use of bridges and routers and/or locator servers that would direct requests from wireless users to these central servers.

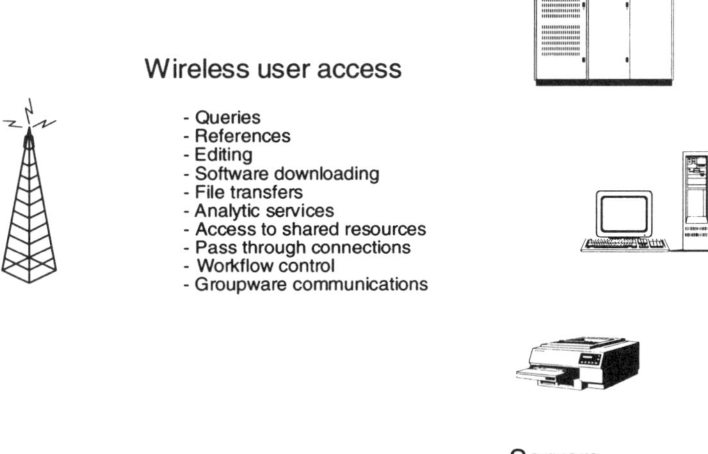

Figure 11.3: Wireless Access to Central Servers

The only difference between accessing central servers versus local or distributed units is that the waiting lines of users may be longer, making it more difficult to assure reasonable response times to the mobile wireless user. This could be improved by assigning priorities to the wireless user to minimize the waiting lines and delays. Other considerations would include caching wireless requests or making them into a delayed call-back type of response.

Locations

The location for making the connection of wireless to wired is very important. The wireless distances are usually the constraining part of the equation, so the location of any interface device will have to be within the distance domain of the wireless world. However, the wired world also has distance limitations varying with the medium being used. If the connection from wireless-to-wired is using 10 Base T cable then the wireless to wired coupler cannot be more than 100 meters (330 feet) from the coupling hub of the wired network. Although these distances are greater than those used in the wireless network, they can still represent severe constraints in the layout of the interconnections.

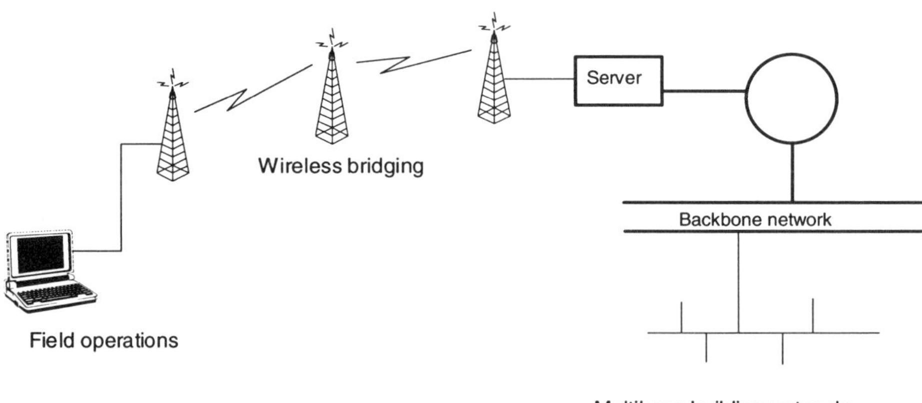

Figure 11.4: Wireless-to-Wire Location Variations

Loads

The coupling between wireless and wired networks will suffer some form of speed imbalance. The wireless network will probably top out at 2 Mb/s, while the wired network will perform at 10 Mb/s and above. This speed imbalance will cause some buildup of load at the connection points. The wired network will have to buffer data to support the slower speeds of the wireless network, and the wireless network will require some form of concentrator to bring its speed up to that of the wired system.

In addition, the loads between the two networks will be subject to the fact that wireless users will use small request packets for the information they need from the wired servers and will likely receive a large number of packets from the server to answer their requests. This will mean a greater high-speed load feeding into a lower speed wireless network. Again, buffering and load balancing will be necessary at the interface point.

Control

The connection between the wireless and the wired networks will require some form of management control. As the wired world is usually treated as the senior environment, control will most likely be vested in this sector. Control will consist of the mechanics to log and manage the processing of a message and/or the management of the request and answer between the client and the server(s).

Control will consist of:

- Message management

- Traffic logs

- Audit logs

- Error logging

- Corruption detection

- Retransmission management

- Other support services

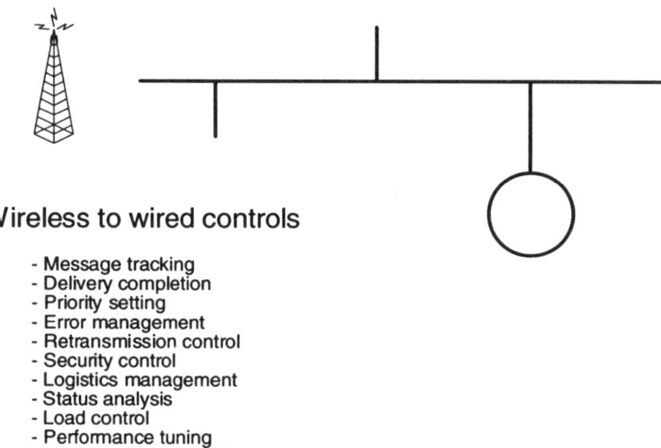

Figure 11.5: Wireless-to-Wire Control

Inbound-outbound transaction management

The movement of messages from wireless to wired will be considered the inbound direction. The wireless network will connect to the mobile clients who will want to obtain services from the wired network. The inbound flow of messages will be the ones that will create the larger load on the system and its services. A few demands from the wireless clients could create major return loads of data to support the mobile client's needs.

Inbound data management can be used to validate and support the planning and control of the overall data traffic. By exploding the level of requests from the inbound traffic, it should be possible to predict and plan for the level of return (outbound) traffic and support some form of intelligent management of this traffic.

LAN-to-LAN operations

The operation of the wireless-to-wired LAN interface will require the merger of two different worlds. Although both worlds are part of the LAN environment, their operations and controls are different. The framing of the data packets may be the same (Ethernet or token ring), but the speeds and the actual operations of the networks will be vastly different.

Recognition

Both the wireless and the wired LANs must be able to recognize the signals from one another. The interface device between the two networks must be able to be a participating member of each network and handle the specific protocol format and signal level needed for each native network. When a signal arrives that is destined to cross over between the two networks, the interconnect device must recognize the request and convert it from one format to another.

Recognition consists of being able to accept the signal in one format and then convert and manage the signal in the second network environment. As the signal reaches the interconnect device, the device must be able to identify, accept, and convert the signal to the proper format.

Acceptance

Once a signal is recognized as moving between the two different types of networks, it is necessary for the receiver to accept the signal as valid. If it finds the signal or message is not up to its requirements, the signal must be rejected and some flag or message sent to confirm the rejection.

The acceptance process will occur at the interface point between the two networks. In a passive coupling the signal will be passed directly to the receiving network and the final terminator or the addressee will have to determine acceptance. In an active network the acceptance would be conducted at the port of entry.

Speed matching

The wire and wireless networks will probably have speed differences. Today the wireless averages 2 Mb/s while the wired networks are seldom less than 10 Mb/s. This difference will increase as the wired networks raise their speed levels to 100 Mb/s.

Speed matching will require considerable buffering to support the return flows from the higher speed wired worlds to the lower speed wireless systems. The inbound matching will be less of a problem.

The speed matching will probably occur at a server type of node which will be able to store the total communications from the wired world and dribble them out to the wireless network. This matching could also provide the active message acceptance process mentioned in the last section.

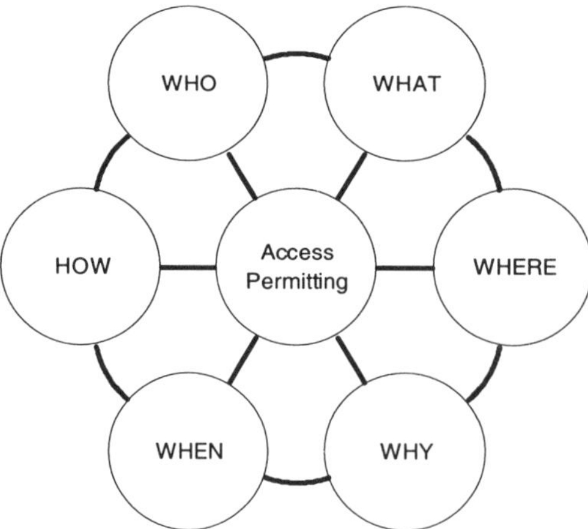

Figure 11.6: Recognition and Acceptance Processes

Packet conversion

Packet conversion is the process of reframing the packets from one network environment into those required by another. Even though the wire and wired networks might both be running Ethernet formats, the detailed headers and packet controls may be different in the two networks. The packet conversion process can be done along with the speed matching and the acceptance process.

Flows

The flow of messages from the wireless and wired worlds will have to be carefully managed and orchestrated. The matching of speed differences, format conversions and audit trail tracking will disrupt the continuity of the flows. The use of message techniques such as electronic mail will provide some form of flow management and control.

The flows between the two network worlds will probably be somewhat discontinuous due to the large variations in capabilities. Buffering or holding the messages to perform the necessary balancing and matching will be part of the design process. Flow control and conversion can be performed by an intelligent server or a large buffer-oriented protocol conversion unit.

Errors

Errors may occur at the point of coupling of the wireless and the wired worlds. Such errors will have a high probability of being errors of interpretation or conversion. If the data are received properly, then the conversion process could be flawed. If the conversion process is made active rather than passive, then the errors should be small and identifiable and correctable before the message is retransmitted.

Losses and reconstitution

There is always the possibility that a message or transmission will be lost in transit. This is possible in any medium and more probable in the wireless world. If the message is missing as defined by a sequential packet identity being open and unreceived, then the system will have to locate and retransmit the missing message.

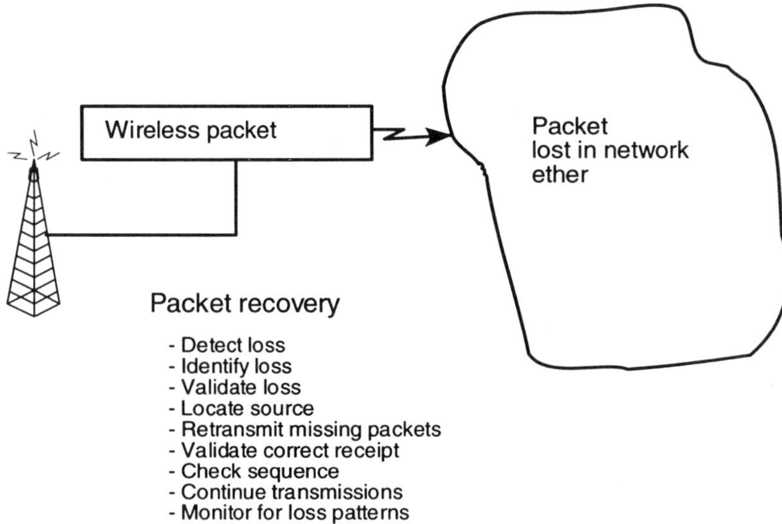

Figure 11.7: Handling a Wireless-Wired Lost Communication

The loss identification and retransmission process can be largely automatic within the send and receive functioning of the system. The receiver will check the message identities and validate the sequential consistency of the received packets. It will also validate the packets to be sure they are of acceptable quality. If any are unacceptable or missing, the receiver will make requests for retransmission.

Cross-couplings

Very few wireless LANs will be independent stand-alone devices. Most of them will be cross-coupled to wired worlds to make up a total integrated (and hopefully seamless) virtual organization. These crossover points will require special attention as there will be a conversion of technologies and signals from the wireless to the wired medium and return.

Cross-couplings may bring together different vendors and different protocols. Any necessary conversions and translations should be done quickly, reliably, and transparently. No user intervention should be required, once the system is set up and thoroughly tested.

Wireless-to-wireless to wired LANS

In some situations the cross-coupling will not be directly from wireless to wired. There may be several levels where one wireless LAN talks to another wireless LAN. For example, in a remote site operation the site wireless LAN could talk to a long-distance wireless LAN and the second wireless LAN could make the connection to the wired world.

Such connections may involve two or more levels of wireless technology with multiple vendors and differing protocols. Automatic connections and end-to-end continuity and error management would need to cover all of the wireless layers.

Multiple couplings

LAN traffic that has to cross over several LANs to reach the intended server or service is involved in the process of multiple couplings. The path could take the packets from a wireless network to another wireless to a wired and back to another wireless LAN. The multiple couplings could involve conversions and bridging to travel over the various layers.

Multiple couplings will increase the overheads of the conversions between the various wireless systems and services. They will also complicate the reliable communication of messages between the calling parties. Multiple connections should be able to be automated and buried the connection management layers of the protocols or the products.

Protocol conversions

Protocol conversions may not be necessary if the wireless-to-wireless-to-wired worlds all run the same type of data packaging. Ethernet or token ring formats should follow the world standards and be used throughout the network layers. There is little justification for adding a new and unique protocol to the wireless layers. However, many organizations use token rings at their wired layers and wish to use Ethernet at the wireless levels. This will necessitate conversion of the message packets just as would have to be done in a wire-to-wire network with the variant protocols.

These conversions are well defined and can be performed in hardware via a multiprotocol bridge or within some part of the hardware coupling logic. Error management and packet control should be an integral part of the conversion logic.

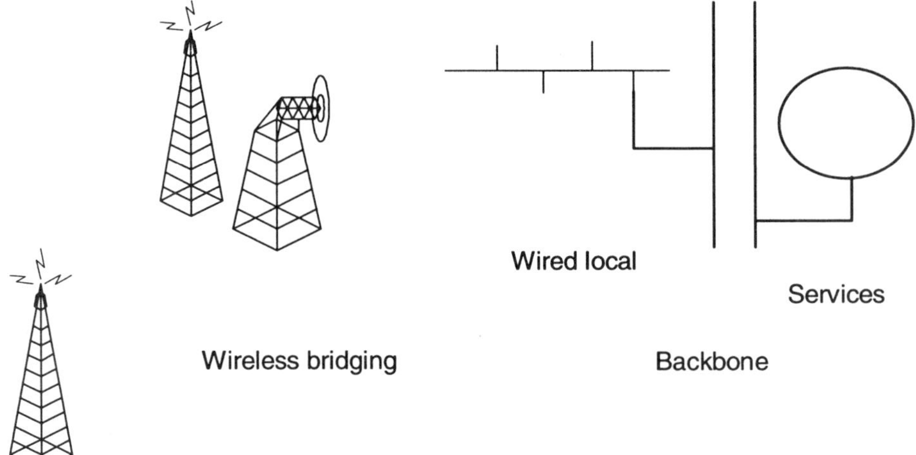

Figure 11.8: Multiple Couplings

Security

The security of message passage over different coupling points between networks and protocol conversions will require careful attention and control. These are points at which taps or external capture could take place. The detail of message frames and contents such as passwords and other details could be exposed at the points of coupling. The use of encryption of passwords and some camouflage of the message framing should be sufficient to maintain the integrity of the security of the total system.

Information transfers

The key component of transfer across the wireless and wired media will be the movement of information of various kinds and types to support the business operations of the users. Most of the traffic will be business data and transactions to and from the mobile users.

The information transferred can be of many varieties. Business operations transactions will likely form the bulk. But queries, requests for remote resources, alarms, communications, messages, and other information can be sent via the wireless and wired LAN worlds.

Service requests

Service requests will be messages that seek to have some other part of the networked resource world respond to a mobile user's needs. Services include:

- Electronic mail

- Transactions

- Database queries

- Remote resource use

- Downloads

- Uploading availability

- Work flow requests

- Electronic forms

Once a service request is sent, somewhere in the system a response will need to be generated. Determining where to address the request and producing and packaging the return messages to satisfy the service request will have to be performed by the network and application layers.

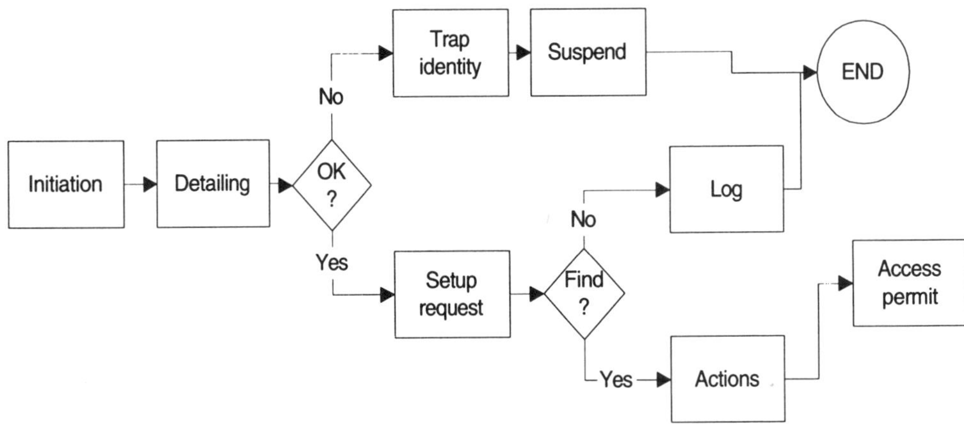

Figure 11.9: Service Request Flows

Resource requests

The request for resources over a multiple coupling network will be performed in the same way as if the resources were on the local loop. The network connecting layers will need to direct the request to the service resources on the proper network link. Resource requests such as printing, faxing, and terminal emulation, can be supplied by any service resource within the network, so the user or initiator does not have to specify the location of the service. Their specification should include the level and type of service and any special requirements such as color printing. The management layer will be responsible for directing the service to an available and capable resource.

Application-based transfers

With application-based transfers, the information on the wireless LAN will be directed to specific server nodes that will be responsible for identifying, accepting, and processing the application activities. The transfer processing steps should be built into the application software and remain transparent to the end user.

- *Transactions*

As transactions are created at the wireless user level, the resident software should be able to determine the need for transfer to a specific location or server. It should create the LAN output package and initiate the transmission without intervention by the user. The transaction level will be responsible for the receipt, editing, and formatting of the application data transfer. It will also be responsible for setting any RSVP requirements and waiting for return notification of the state of the transaction and its remote processing.

- *Data requirements*

Some application-based transactions will require access to additional remote data to complete the local portion of the transaction. For example, in an ordering application the remote order unit will need to determine product availability from a remote inventory server and possible the price from a pricing table or product master file. Both of these requirements would be met by the user/client unit sending request messages over the network for the data needed to complete the local portion of the transaction.

Upon return the local software would continue processing and complete the entry of the transaction.

Lookups

In other transaction applications, especially those dealing with requests and queries by outsiders (kiosks, sales, proposal writing, etc.), the local unit needs access to remote server databases to answer the local query. The request could be a customer inquiring about the stock availability of a product, a salesperson requesting a price check, a marketing representative working on details of a proposal, etc. Each of these transaction sequences requires the lookup and return of information from a server database. The local client station would frame and send the request and the server world would need to process and respond to the request in a timely manner.

These lookups may need to travel over several layers of the network and visit several servers until the requested information is located and packaged for the return to the initiator. The dissemination of the lookup and its return to the initiator should be transparent to the user.

- *Authentication*

The process of approving and authenticating the legitimacy of a request or a transaction should also be made a transparent part of the application process. The transaction can be labeled with the user's identity and encrypted passwords. The servers or the network management system should be responsible for checking and validation of the user's and their rights and permits to perform the requested transaction. The final authentication on database access to record and field level would be allocated to the database management system.

In the event that the transaction is not authenticated, a return denial message would be created and sent to the initiating user for review and further action. A recreation of the message with new facts or release codes could be created, or some external management steps and decision processes invoked to resolve the authentication conflict would have to take place.

Work flows

Work flow processing involves directing transactions from user to user for action and decision making that follows an established business procedure. An input that is created by a wireless user might need to be reviewed and approved by a user on a wired LAN. The work flow process would identify the need for the approval, define who would have to do what to complete the approval, and route the transaction from the originator to the approver with an attached message requesting the approval step. Once the transaction is initiated, the work flow process would be responsible for transmission of the request and for the follow-up to see that it is completed by the receiving user.

Work flow is based on the use of electronic forms or electronic mail messages that require action and response from other parties in the organization. As the wireless layer of the LANs is most likely to be at the lower data creation layers of the organization, the use of work flow to define and direct the transaction flow may be a logical choice. Tracking the messages and overseeing their routing and completion would fall to the work flow controller, which could be located at the initiator level or an intermediate server. In either location, the work flow controller would oversee the movement of the requests and the transactions from one node to another and keep track of the state of the transactions at all times.

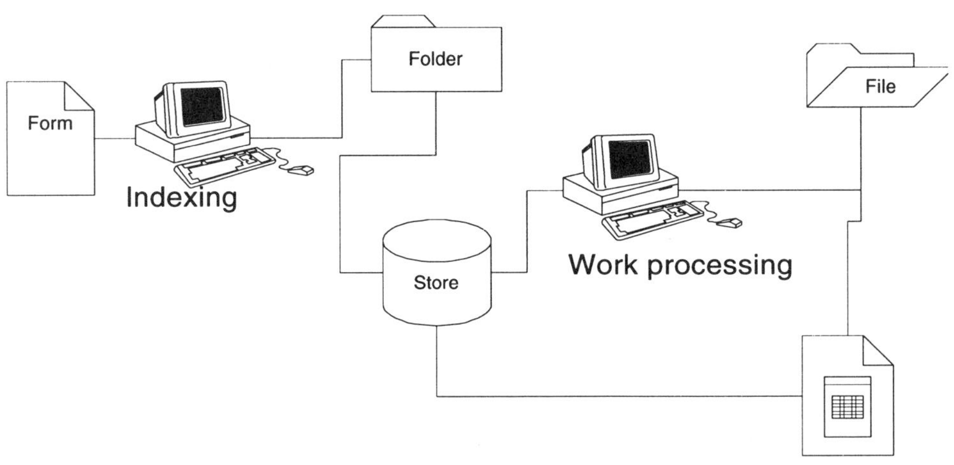

Figure 11.10: Work Flow Processing

Work flow provides information procedures for improving the overall end-to-end processing of transactions and communications of an organization. The work flow steps can be defined by type of transaction and implemented as intelligent agents when that transaction is initiated. Because work flows can move throughout an organization to achieve their completion, they will be actively involved in cross-couplings between wireless and wired LAN worlds.

Messages

The sending of messages within an organization can also involve considerable cross-coupling traffic between wireless and wired LAN worlds. Electronic mail, memorandum distribution, action items, notes, reports, and other forms of communication documentation can be created or disseminated to and from the wireless LAN units. These messages will require routing, tracking, backup, and archiving.

Messages often generate return or extension messages and can become parts of other files and collections. The passing of messages between the user layers, including the cross-coupling of wireless and wired LANs, can account for as much as a majority of the traffic on the networks.

User messaging

The wireless and wired LANs will work together to support user messaging and electronic communication. Users will create messages and pass them to their local LAN. The messages will need to find the addressed user or a server database that stores and mailboxes messages for the intended recipient.

User messaging can include open text messages and also transactions and data lookup messages. The messages should pass from wired to wireless to wired and back again without specific involvement of or concern by the users.

Messaging will probably be a high percentage of the communications that are handled by the wireless LAN. Many of the systems are oriented to transaction processing and will package the transactions as messages for two-way communications between the servers and the client user stations.

Electronic mail

Electronic mail represents a large portion of the potential data that will be transferred between wireless and wired worlds. Users can package most requests and communications messages into E-mail formats and use the E-mail application services to send and manage their communications processes.

The use of electronic mail as the transfer service will provide many administrative support services, such as:

- Logging

- Audit trails

- Arrival notices

- Access flexibility

- RSVP

- Return addressing and transmission

- Mailboxes   - Mailing list management

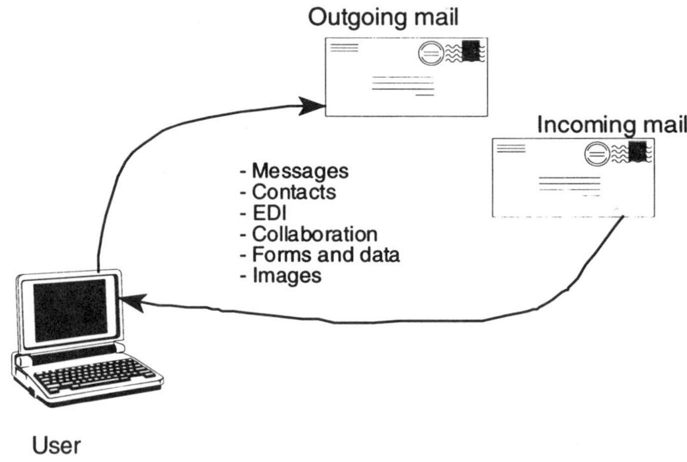

Figure 11.11: Wireless LAN E-Mail Services

Comments and notes

Comments and notes sent via the wireless LANs are a subpackage of the electronic mail process. Notes and comments are usually attached to someone else's electronic mail package and can be read and processed as separate but coupled messages.

Comments and notes may be collected from several resources and applied to a single basic communication product such as a report or presentation. They will usually require some form of review and possible action on the part of the original author(s) of the communications package.

The comments and notes will need to be accessed, evaluated, and answered as part of the groupware communications process. Products and tools are available to assist the communicators in performing these processes. The network managers would have to oversee the movement of the messages between the parties, including the movement over wireless and wired worlds.

Interactions

More and more applications are moving to support interactive messaging and responding as part of their architecture. This will increase the need for active, integrated networks to take the back-and-forth message traffic between the parties. These interactions are also focusing more on the active work of the organization and securing the inputs of concerned parties to improve the quality, timeliness, and the delivery of the goods and services of the organization.

As more and more businesses reengineer themselves into flat, just-in-time production and delivery units, the level of interactive messages and transactions that will be managed and tracked through the organization will grow. The networks must be capable of seamless, high-speed support of these interactive processes. The interactions will involve individuals and layers of the organization. They will occur on an event-action-driven basis and generate response requirements that will be actively tracked and managed via the network.

Voice-video packets

As information becomes more interactive, the use of just text or images will become a limiting factor of the communications process. By adding voice packets to explain the message with verbal information, the receiver and the sender can link up another set of parallel senses to support the message processing.

Another sensory process can be added to the information interface by including video sequences or live video-teleconferencing in the message. The addition of voice and video to the message stream will require time synchronization of the message flows and, in the case of video, the demand for high levels of speed and time ordering to maintain the 30 frames a second display rate of full motion video.

After these levels are integrated so that they can move from wireless to wired and back again, the technologist will add to the information circuits the last two major senses, namely smell and touch. Touch can be done via virtual reality concepts and smell will require some new distribution and packaging mechanics.

Roaming

Many wireless systems are flexible enough to allow end users to be completely mobile within their wireless LAN coverage space. This is accomplished by using roaming to couple the mobile user to the wireless resources. In addition ,a user may move from one wireless LAN area to another and still achieve connectivity. This level of roaming is similar to that used in cellular telephone systems when users leave their home service territory and seek service from another area.

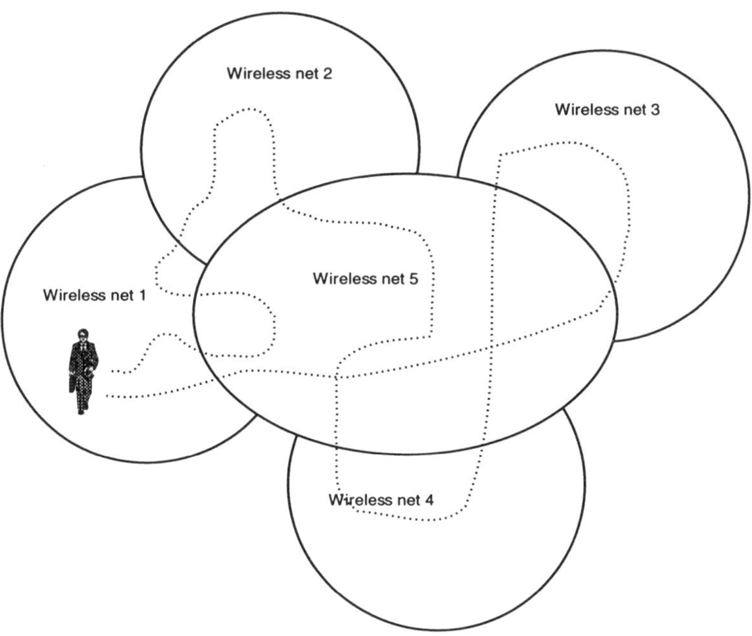

Figure 11.12: Wireless Roaming

Roaming adds a new level of complexity to the management of the user access and services of a wireless LAN world. As users move to different locations they can pop up on different segments or entirely different wireless networks. There will be some form of central registry server that can identify, verify, and link the users back to their available resources and services. In addition, the system will have to maintain a log of current operations so that a user who disappears from one wireless world can be picked up on another and have the services continue as though they were continuous and noninterrupted.

Roaming is a convenience for the mobile end user of wireless LANs; however, it is a major technological and operational nightmare that will need to be resolved within the specifications and components of the systems. Wireless LAN roaming uses some of the concepts proven in the cellular telephone system, which has the system negotiate with an adjacent cell for a takeover of the call and maintains continuous connectivity as a user reaches the edge of a cell, The use of the cellular tracking service that knows where all active users are at all times is also appropriate in a mobile wireless LAN environment.

Bulletin boards

The use of bulletin board systems for user messaging will also affect the wireless cross-couplings. The bulletin board would logically be located at a convenient place where all users could gain access to its contents and services. The wired LAN servers would be most appropriate for the bulletin boards, as they are served by faster communication connections and reach a broader audience. The wireless units would reach the wired bulletin board servers for the posting or extraction of messages and information.

The bulletin board concept can be used as a trading post for messages, transactions, and other interlinked information flows of the organization. Security and access control will need to be carefully managed, but the bulletin board concept of information transfer is simple, inexpensive, and easy to operate.

Interlink issues

The interlinking of wireless LANs and other worlds of communications and data services will require a major technology evolution and be a critical design issue for any wireless LAN implementation. Interlink issues will need to be resolved by the industry, within individual applications and within the user organizations. These issues will change as the various components of the connecting levels change. For example, the move to 100 Mb/s wired LANs or the introduction of ATM to the desktop will have major impacts on the evolution and development of wireless LANs. As these technologies develop, the wireless LANs will have to interlink to them in order to continue to provide their services.

Standards

The development of standards for interlinking will be one area of coverage for the interlinkage issues. The problem today is that the wireless LAN world is fragmented across many vendors and technologies with no consolidated organization able to represent or promote integrated standards. For the foreseeable future the standards will likely be set by the wired LAN world and the wireless world will find a convenient hole through which to couple its services. By coupling the wireless world to wired servers using dedicated or vendor proprietary standards, interlinkage will be accomplished by piggybacking on the wired world of standards.

Conversions

Conversion of the signals, protocols, and packets may be required in the interlinkage of wireless and wired LANs. Any conversions should be accomplished with as much transparency as possible. All of the data levels of the systems should conform to a defined standard. The conversion should be between known and defined entities and therefore be amenable to implementation at hardware or firmware levels of the system. This would remove the conversion process from the view and concern of the users.

Losses

In the interlinking of LANs there is the possibility of loss of the data packet. The loss could be caused by the moving beyond the distance limits of the wireless world or by interference at the interconnection point. In any event, the loss of data in the LAN world is a potential problem. If the systems can provide some form of local backup for transmissions, then as soon as the interconnection was reestablished the units could retransmit the missing data elements.

Loads

The loads at the interlinkage points can be a major operational problem for the performance of the overall system. Because all of the wireless units are likely to travel through one or a few interconnection points, the loads will be concentrated at one point. The tracking and managing of the loads to keep them under reasonable levels will be a systems management function. If the loads are too high, shedding of low-priority work and/or introduction of alternate service routes would be warranted.

Wireless-to-wired LAN management

The point or points where wireless and wired LANs meet will need to be supported with some special management considerations. The signals will be traveling from one medium to another and taking on different form. The management connection will have to:

* Recognize the various types of signal traffic

* Validate the traffic

* Authenticate the security

* Accept and transfer the messages

* Log the messages

* Perform accounting services

* Provide error management and retransmission if needed

* Supply fault management and recovery services

The interfacing between wireless and wired LANs should be made a part of the operational management architecture. User's should see the LAN world, both wireless and wired, as a transparent continuum that provides them access to their information resources and services. The management steps should be built into the software and hardware that bond the networks together. The traffic and information flows should support the applications and the user requirements, not the demands of the technologies.

Extendability

The wireless-to-wired LAN connections will provide a mixed technology extendability to user and systems services that can be efficiently and costeffectively supplied by a combination of technologies that complement one another. In most systems the wireless LAN will not compete with or displace the wired world. Each has a place in the scheme of opportunities and applications for network systems and services. The wireless world can extend the wired serviced to mobile workers and to applications that are not well served by fixed wires. In addition, the wireless LANs can supply interconnection between certain remote LANs better and cheaper than by using third-party services.

The wired world still dominates the information services and database worlds. By coupling the wireless and wired worlds together, both will benefit along with the user and the organizations that choose to use these technologies.

CHAPTER 12

Long-distance Wireless LAN Systems and Services

Wireless LANs are usually thought of as being localized, limited-distance systems. However, there are several versions of wireless technology that can provide long-distance wireless connectivity between LANs. These technologies are all wireless and operate in similar fashion to the localized wireless LANs. However, some of them can extend the connecting distances to several miles and others can literally link systems around the globe. This chapter will cover the long-distance wireless LAN services.

The short-distance wireless LANs are mostly all within a building envelope. These are known as WINs or wireless inside networks. The external networks are knows as WONs or wireless outside networks. The WINs are limited distance, localized systems and the WONs are longer distance (mostly limited to a line of sight of approximately

15 miles) metropolitan area networks. By coupling these two forms of wireless LANs together, it is possible to cover a wider area and provide coverage to a multiple building campus environment.

LAN-LAN hopping

LAN-to-LAN hopping involves a message generated on one LAN being able to move onto other LANs for delivery to its final address destination. As more and more LANs are interconnected, the hopping concept become the vehicle for making the overall interconnectivity seamless and contiguous. In long-distance wireless LANs, the hopping concept would involve the coverage of the longer open segments of the network and the move from a WIN to a WON to another WIN and finally perhaps to a wired service.

LAN-to-LAN hopping should be performed as an automated service that is seamless and reliable. The use of intelligent bridges and routers, which can integrate the different network technologies, is the key link in this level of enterprise data movement.

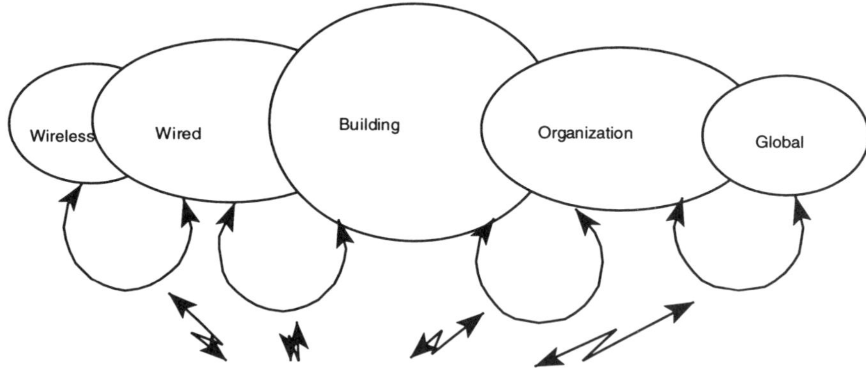

LAN - LAN message hopping

Figure 12.1: LAN-LAN Hopping

Interlinking

The interlinking mechanics will involve the ability of a LAN to recognize signals that are destined for another LAN world. Once recognition is made, the message would need to move to the interlinkage point and flow smoothly from the source LAN over the interlinkage onward toward its ultimate destination.

Interlinking requires access to the level 3 or 4 OSI information which defines the network and transport layers or alternately to the TCP and the IP contents. With this information the interlinking process can determine the movement and possibly the routing of the message to its proper destination.

Integration

The use of short-and-long-distance wireless LAN services can support the efficient and reliable integration of information flows across the enterprise. The integration will take place through LAN-to-LAN interfaces and the merger of wireless and wired network technologies. The integration will also be supported by the abilities of the information management process to direct messages and user requests to the proper service locations.

The overall concept will be to build an integrated network services world that can traverse different technologies and provide responsive user services. The integration level should be focused more on the provisioning of user services with the integration of the technology layers handled as a seamless and transparent background process.

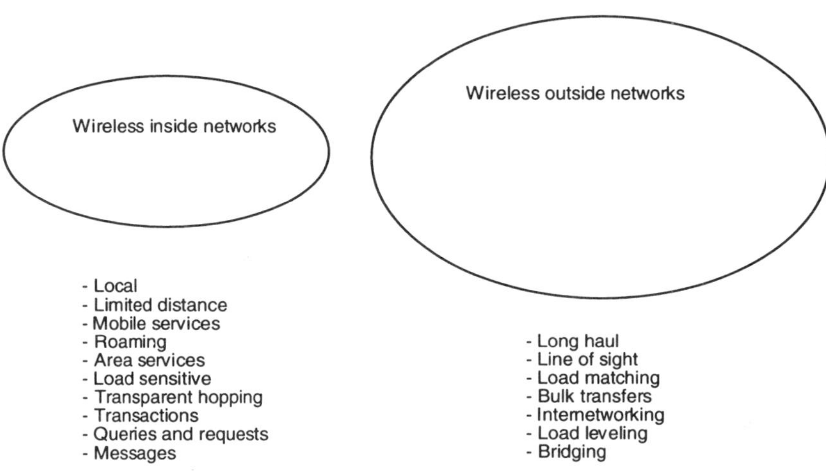

Figure 12.2: WIN and WON Variations

Local to global

As wireless technology moves from local limited services into providing longer and longer distance services, it will become a major link in the overall information services spectrum. With the growing use of satellites and the demands for anywhere, anytime communications couplings, the wireless world could become the dominating technology in the overall equation.

The key is that wireless connectivity can be local, metropolitan, wide-area, and global in form. Although the technologies vary, the wireless implementation provides flexible, untethered couplings between the users and their service systems. The fact that wireless can go anywhere and provide a broad range of services makes it a logical building block for an enterprise information environment.

Mixing of technologies

Most enterprise networks will be an intermix of various technologies. The mix will range from various forms of wireless to different types of wired to a variety of wide-area forms. The mixing will provide the unique services and coverage of each type of technology to support the needs and applications of the organization. The network management system and services will be responsible for the conversion, translation, speed matching, and other interconnectivity services. To the end user the intermix of technologies should be transparent, efficient, reliable, and consistent.

Cellular connections

Cellular services are based on wireless radio wave networking. Primarily oriented to voice communications and built on analog waveforms, the cellular services networks provide limited-distance cell-to-cell connectivity and interfacing to the wireline communications world. Like voice/analog networks, the cellular system can be used with analog modems to transmit and receive user data communications. If the remote cellular users connect to a LAN and use LAN services from a remote client computer (such as a laptop), they are a part of the wireless LAN services environment.

Cellular connections can be treated as a special remote extension of wireless LANs. They can provide limited-speed connectivity and data services to a variety of remote users.

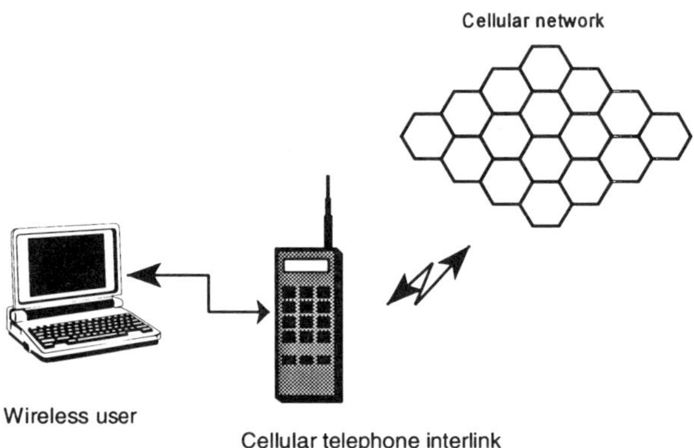

Figure 12.3: WIN-to-Cellular Connection

Unlimited spatial coverage

The key to cellular services for connecting to a remote LAN is that they are relatively distance insensitive. There is very complete cellular coverage of most modern population centers and the major transportation corridors. In the United States there is also good cellular coverage of rural areas, where farm and service managers are high-level users of cellular services.

With the roaming services (and the costs associated with roaming), users can have a broad spatial coverage for their travels and still maintain some form of contact with their remote data services environments.

Open air services

Cellular service coverage can extend from transportation facilities to the open spaces. Remote field sites, vacation spots, and walk-in locations (such as remote work sites) can usually be covered with cellular services. This means that wireless contact can be maintained with LAN data systems from almost anywhere via the cellular services.

Radio links

It is possible to build and maintain remote radio wave services for connecting remote users to their organizational networks. These links will be built around the Specialized Mobile Dispatch Radio (SMDR) systems that have been used for years by utilities, construction companies, taxi fleets, and other mobile services units. SMDR is now being converted from local service to an integrated national network with higher speeds and quality for the movement of data between remote users and their data systems. Nextel, Inc. is one of the organizations leading this adaptation of older radio technology to wireless data systems use.

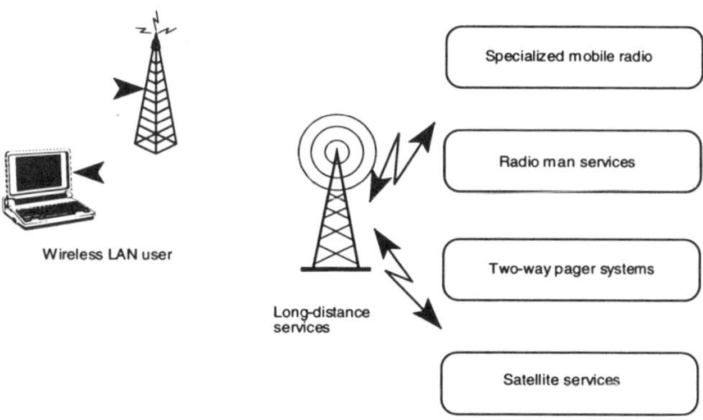

Figure 12.4: Long-Distance Air Wave Services

Wireline link services

Most of the cellular and radio link services will connect to traditional wireline services at some point. Some users may contact other cellular users for local client-to-client data transfer, but most users will be contacting a fixed organization location for most of their data services. This means they will be interfacing to plain old telephone services (POTS) for some portion of their routing and traffic movement.

Wireline services will convert the data signals to their modulated format and then move the data through their network routing system to the addressed location. The wireline data services will be readily available, reasonable in cost, slow in speed, and moderately reliable in their delivery of the data.

Cellular digital packet data

Cellular Digital Packet Data (CDPD) is a new format for packaging digital data for transmission over the analog cellular network. The messages still travel over an analog network, but they are packaged and managed in a digital format. This will increase the reliability, error control, and signal management within the existing cellular systems capabilities.

Digital cellular

Cellular services are now moving from their analog base to a full digital signal process. This will provide more channels of service and greatly improve the reliability and management of the signal flow processes. The all-digital cellular service will also make it easier to transfer data between cellular users and their remote LAN environments.

Personal Communications Services (PCS)

The new of personal communications services (PCS) may have a significant impact on the wireless LAN world. PCS is a low-power, localized cell service that is intended for person (rather than vehicle) mobility. This is the same market covered by wireless LANs. The major difference is that PCS will allow roaming in a wider geographic area than those provided by the wireless LAN. PCS will also be a commercial carrier service rather than a user-owned service.

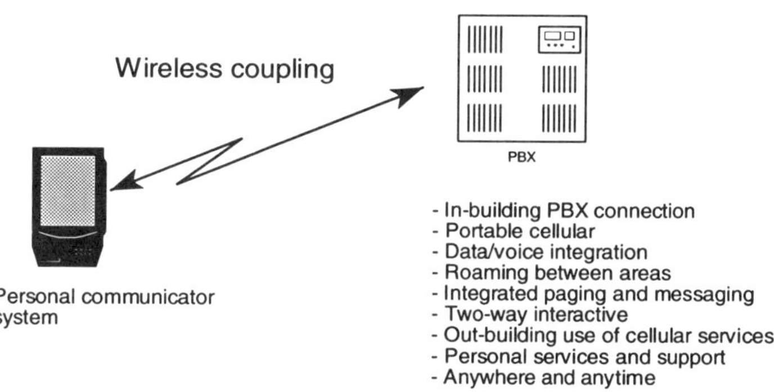

Figure 12.5: PCS for Wireless LANs

Personal communications services may be offered as packages that can be implemented within building envelopes. One option would be to connect a PCS server to the internal Private Area Branch Exchange (PABX/PBX) and allow roaming users to be able to couple to any internal or external telecommunications service. Such a coupling would be directly competitive with or complimentary to the concepts of wireless LANs.

New frequencies and power levels

The PCS will operate in a newly allocated range of frequencies in the gigahertz bands. The frequencies are radio frequency and capable of both voice and data traffic. These frequencies were previously used by emergency services for reliable local communications.

The PCS power levels will be very low to allow limited coverage of the areas without interference with other nearby cells using the same frequencies. Cell hopping and roaming will be provided in a fashion similar to that in use within cellular systems. The PCS will be all digital from the start, thus avoiding the conversion from analog to digital that the cellular world is now going through.

New PCS services

The key service from the PCS products will be a single communications device with one identifying number that works on everything at any location. The concept of a singular truly universal telephony connection is very attractive in today's convenience-oriented market.

The new PCS will also provide integrated services for voice and data communications and may eventually be expanded to video. Remember the Dick Tracy wrist communicator that handled radio, video, and messages? Well, PCS is the arrival of that old comic strip prognostication. A truly portable, go anywhere, connect anytime communications system would provide an efficient and easy-to-use interface that would depend on internal network services to direct and handle the provisioning of the data, message, and information deliveries. PCS may become the universal communications appliance. Many firms are bidding for the air wave space to run PCS systems in the hope that they will become the wireless communications base for the future.

In-building PCS

Personal communicator systems are normally considered a low power, out-of-building communicator service. If the PCS cells are set up so they are within a building envelope, then you have an in-building PCS service. Such a service could provide wireless connectivity to internal users from any location within the PCS coverage space. As the PCS is within the building envelope, it would not need licensing and it would be the dedicated property of the organization (ie., no monthly telephone bills).

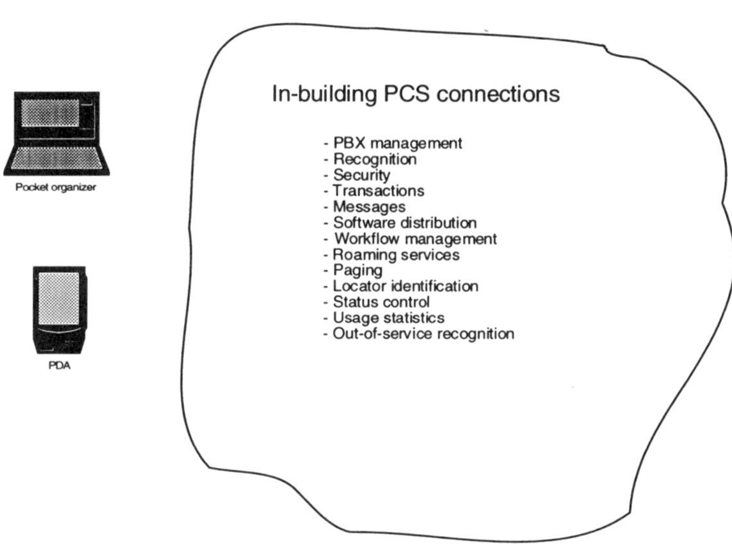

Figure 12.6: In-building PCS Connections

The in-building PCS would allow users to take their desk telephones with them throughout the building. They would be reachable by normal call processes anywhere they roamed within the covered space. PCS would replace the desktop telephone and all of the in-wall wiring.

The PCS service can also handle data, so it could interface to portable personal digital assistants and provide another wireless data link. This could make dual-function devices like the Bell South Simon product, the telephone/PDA of choice in many organizations.

PBX connections

The private area branch exchange, PABX or PBX, can be a logical coupling unit for wireless LANs. The PBX is the connector between internal and external networks. Longer distance wireless LANs will need to interface between the internal service networks and the longer distance worlds. This is the same type of connection being managed by the PBX in the wired world. Extending the PBX to provide interconnection to long-distance wireless LAN services will use standard services of the PBX and provide support to a broad range of users.

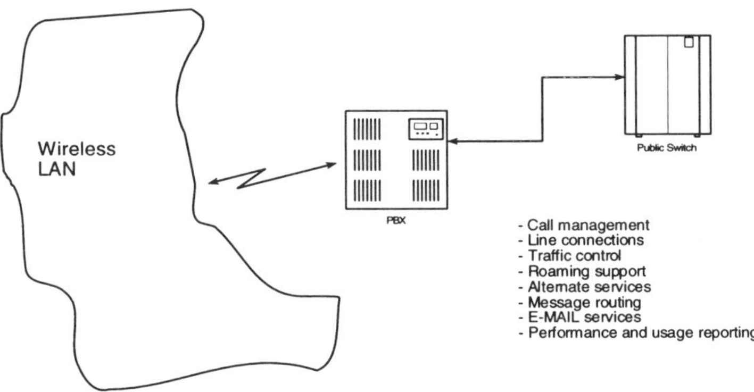

Figure 12.7: PBX Connectivity for Wireless LANs

Low-power roaming

Low-power roaming of mobile users involves the use of personal communicator products that can make connections to local PBXs within building envelopes. The low-power systems will provide in-building connections to networks and data servers. The user can roam around the building and make calling connections to the LANs and data servers. Data and message services can be moved from the networks to and from the user mobile devices.

By staying at low power, the units should not require licensing and can be used as owner-operated networks without time service contracts. The low-power roaming units should provide low-cost, low-volume data interfacing services.

Digital PCS

The PCS will open up new bandwidth communication channels that will support a wide range of wireless data services, including connections to wireless LANs. PCS will provide low-power, cellular-type services using digital channels.

Low Earth Orbiting Satellites (LEOs)

One of the new distance services for LAN-to-LAN interconnection is the use of satellites traveling in low earth orbits to receive upward transmissions, hop them between other satellites in the same orbit, and then beam (downlink) the signals back to earth and the receiving system. Known as LEOs, these projects will provide low-cost , high-performance digital communications services that are distance insensitive.

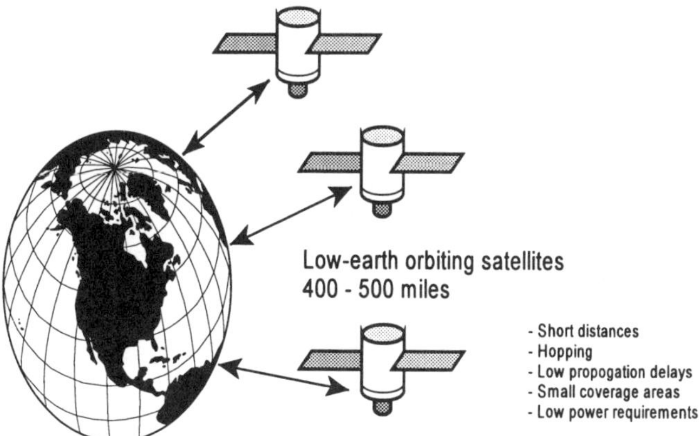

Figure 12.8: LEO Configuration

LEOs will not only provide wireless linkages to other networks and services, they will also strive to be worldwide. This is not fully possible, but they will band the earth with their satellite coverage and be able to hit any 3000 mile wide space in either half of the globe, north orsSouth, depending on where the satellites are placed.

The LEOs will offer a broad range of services that will include anywhere, anytime access by users to their larger data networks. Also to be offered will be voice and video communication services plus news updating and other communications services. The concept is a general communications network that puts the user in contact from any point within the near-global service view of the satellites.

Irridium project

The Irridium project is a joint venture of Motorola and General Electric to place upward of 66 satellites into orbit 477 miles above the earth's surface. Spacing the satellites about 1½ degrees apart (as measured from the earth's surface) the Irridium will provide high capacity, good performance, interactive services to users for a wide variety of applications.

The Irridium project is now being funded (3.4 billion U.S. dollars) and expects to begin satellite launching in 1996. Services would come on line in 1998.

The project has taken considerable time to germinate and receive approval from various governments. It is still not a completed deal, but the technology and the uses appear to be well founded and the needs for such services are growing beyond the capabilities of existing systems and services to support them.

Politics has played a large part of the launch of the Irridium project. The original proposal included the use of space switching, satellite hopping, and terrestrial carrier bypass. The terrestrial bypass (which would cut out local carriers from the revenue loop) has now been dropped as a political concession.

Odyssey

The Odyssey project is sponsored by Hughes Aircraft Corporation, a subsidiary of General Motors Corporation. Odyssey plans to put up 77 satellites in a orbit of 450 miles. It is a similar to and competes with the Motorola Irridium project. It basically blesses the concepts of the LEOs and shows that there are entrepreneurial monies and large corporations willing to back major global projects in the communications area.

Linkage

The linkage in LEO projects is wireless from the base level user environment up to the local satellite that is passing overhead at the time of transmission, wireless from satellite to satellite to satellite until the signal is over the receiving territory, and then wireless down to the addressed receiver.

Connections

The connections from ground-level users to the LEO environment will be through channel access contention. The requesting users will signal their LEO interface that they need to communicate over the satellite loop. The interface will arrange for circuit bandwidth and make the connection to the overhead LEO satellite using an open channel. Once the communications packet is released to the LEO environment, the service will become a datagram routing process with the LEO managing the setup and movement of the communications packets between LEO satellites and the eventual ground receiver.

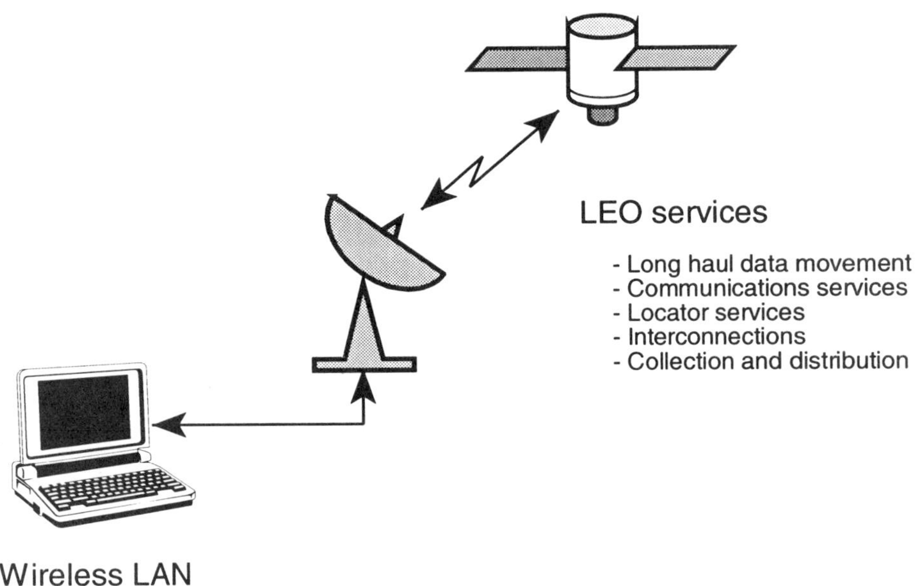

Figure 12.9: Wireless LAN and LEO Services

Once the traffic has moved to the receiving LEO zone, the local LEO will contact the receiver's ground interface and download the communications packets into its buffer memory. Movement to the internal computer systems will be by awaiting message notices.

Satellite hopping and space switching

As the communications packets move across the LEO world they will be handled by a sophisticated satellite hopping and space management system. The switch has to know the destination of each packet and compute the correct positioning satellite for the downlinking based on time to travel the hops to the correct receiving location. Errors, retransmissions, load buffering and other traffic must be taken into consideration. Depending on the number of parallel satellite arrays, there is an opportunity to use alternate routing for the messages. However, if there is only one planar sequence of satellites, the traffic would have to hop in order to get from one satellite to its next neighbor.

The satellite hopping and space switching process will be the most complex part of the LEO operations. These processes will have to take place up in the satellites and their control systems. If they fail or cause errors, they will be difficult to repair. Backups and tandem processes will be a critical part of the overall LEO configuration. Ground control and management of the overall traffic will also be very important.

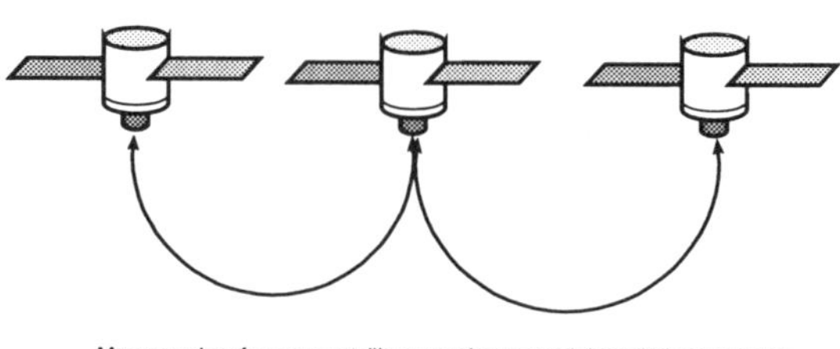

Figure 12.10: Satellite Hopping

Global services

One of the advantages of the LEO system is that it will have to provide global service within the apogee of its satellites. This means that the costs of sending a message across the street or around the world are nearly the same. With the wireless LEO system much of the external WAN distribution of data and information (non-time-dependent transmissions) could move to this service and be distributed by the LEO system anywhere within its worldwide coverage area. Multidrop communications would likely become a speciality of the LEO systems.

State of the application

LEO systems are evolving as a hot property. They are being supported by many large organizations and governments. All of the concepts needed for the LEO systems are in use in other systems. Cell hopping is similar to satellite hopping, other orbiting satellites can control the sending and receiving of high speed communications, frame relay is a protocol similar to the one proposed for LEOs, and the network management for ground satellite interfacing exists in many forms.

What will need to be proved with the LEOs is the market acceptance, reliability of services, and cost competitiveness with alternative forms of traffic. The significance of the organizations backing this approach and the amount of investment being collected probably make this a more sure bet thanthe introduction of cellular telephony was ten years ago. Time will tell.

Medium Earth Orbit Satelllites (MEOs)

MEOs are a close cousin to the LEOs, except that they are placed into a higher earth his allows fewer satellites to cover more surface space. However, the trade-off is that each satellite will have to handle more traffic and the timings to and from the higher orbit will need to be factored into the transmission equation. It will also cost more to launch the higher orbit satellites, but the trade-off is that there will be fewer of them.

The satellite hopping and space switching process will be similar to that of the LEOs. MEOs will also offer communications and interconnect services on a global basis.

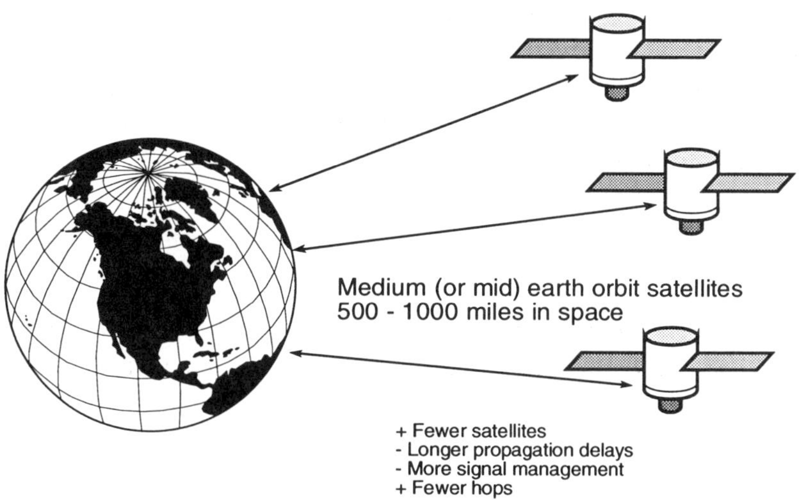

Figure 12.11: MEO Operations

Odyssey project

The Odyssey project involves another consortium of different vendors and suppliers. It is headed by TRW and Teleglobe Canada. Their plan is to place 12 satellites in orbit and offer voice, paging, data, messaging, and positioning services. The total investment is expected to exceed $2 billion.

GlobalStar

GlobalStar proposes to operate 48 MEOs for worldwide services, similar to Odyssey. This project is a partnership of Loral Communications, Qualcom, Alcatel, Alenia, France Telecom, DACOM of Korea, Deutsche Aerospace, Hyundai, and Vodaphone Australia. This is a worldwide consortium but such a diverse group that its meeting must appear to be a little United Nations. Estimated investment is $1.8 billion with operational plans to go live in 1999.

Elipso

Elipso is another consortium consisting of Westinghouse, InterDigital, Mantra, Harris, and Barclay's Bank. They plan to use 16 to 18 satellites placed in an elliptical orbit. The orbits would cover more of the high-density population centers of the world and provide more traffic potential with less satellites.

Teledisc project

The Teledisc project is the brainchild of Microsoft chairman Bill Gates and McCaw Cellular's chairman Craig McCaw. It proposes to set up 840 multilayer satellites covering all of the globe's surface and offer universal, low-cost, broadband data communications services. The proposed total costs exceed $14 billion. This is probably the most ambitious of the LEO/MEO projects, but it has the world's most experienced technology entrepreneurs at the helm and may have the muscle to become a reality in the future.

VSAT satellites

The very small aperture satellites (VSATs), which use small 18-to 30-inch dishes on the ground are another attractive form of wireless LAN communications. These will use high-altitude geostationary satellites and provide a range of medium and high-speed interconnectivity services.

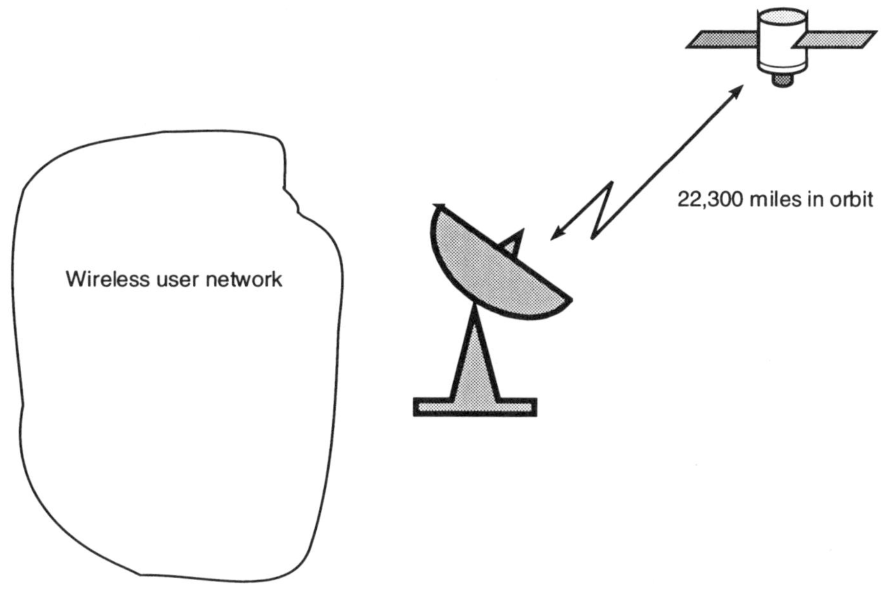

Figure 12.12: VSAT Connections

Locational connections

The coupling to the VSAT network is via a tunable small satellite dish. Such units can be portable and battery operated, providing some level of flexibility in setting up and operating a wireless local LAN and using the VSAT to interconnect to the rest of the world. The military successfully used this concept in the Desert Storm engagement and plans to increase the information couplings via VSAT for its mobile deployment forces.

Physically remote sites

The LEO and MEO products and services will provide an extended range of coverage for physically remote sites. Depending on their placement, these systems can cover large sections of the globe with their services. Being moderate speed wireless services, they can provide remote users with access to a complete communications, messaging, and data world.

Physically remote sites will now be able to be connected to other parts of the organization and share services with the rest of the world. These sites can access data, provide messaging, input transactions, perform communications deliveries, and be queried by other parts of the organization.

Portable dishes

The connection of satellites to remote locations will be done via portable satellite dishes. The VSAT dishes are in the 18-to 30-inch size. Such dishes can be moved about easily and assembled and focused to the satellite transponder in a few minutes. Once set up, the VSAT portable dishes can provide a high-speed intercommunications services.

Global coverage

The use of satellites and VSAT technology can provide global coverage of interconnection to remote information services and resources. This means that a user can move to any spot on the globe and set up a connection to the satellite system. From the satellite service, users can move information and data to and from their location to the processing points.

Global coverage will increase the flexibility and mobility of information systems users. They will be able to make contacts and interfaces from any mobile position. Although the systems will require physical set up through the satellite dish, they will provide high-speed, reliable data transfer services.

Portability factors

The portability factors of VSAT systems will involve the mobility of the VSAT dish and the availability of adequate space to set up and tune the dish and satellite alignment. Portability will also be influenced by air space interference, tuning alignment, and signal quality.

If the satellite service cannot be quickly and reliably set up, the users will be denied access to their information resources. Speedy set up and easy tuning of the dish to satellite alignment will be key to portable connectivity.

Mission support

The mobility of portable satellite systems will allow them to be used in support of remote projects and missions. Situations such as field operations, emergency efforts, remote project sites, military operations, and other missions can be serviced by these systems. Their support will allow these remote missions to maintain connectivity with their command centers, transfer data and messages, exchange files, and maintain data communications.

Security considerations

The satellite implementations of wireless LANs will need to pay special attention to the security of the data transfers. The satellite services cover large distances using open space that can be trapped and scanned by technically astute individuals and organizations. Most satellite messages will need to be encrypted to allow message handling with reliable security.

Other wireless services

A number of other wireless communications services could be coupled to the wireless LANs to make for a national or global information system. These range from specialized mobile radio (SMRs) which is the local dispatch radio service, to the national pager networks and air telephones. This section will provide a quick review of these technologies which may be used in combination with wireless LANs to extend their service relationships.

Air link messaging

Air link or radio paging services can provide for the intertransfer of data and communications messages. These services such as skypage will allow users to make contact with service centers and with originating organizations via the wireless service. The air link services can be a local and remote form of wireless interconnection than can support remote computers and LAN servers.

The air link services will be at low-level communications speeds, but they can be widely available and easy to use. Their costs will also be low and based on a message volume measurement.

Pager services

Pager services can also be used to move data between users and their organizations. Although intended for short text messages, the pager networks can be extended to support longer messages and some file transfers. They can be used as a temporary data mover at low speeds or as a remote connector on an occasional basis.

Pager-to-LAN connections will be used for low-level applications with limited volumes of information movement. They can support data collection and limited queries and accesses.

Air phone

The telephones that are being placed in airplanes can also be used for the interconnection and transfer of data between flying users and their organizational data systems. Air phones will operate using terminal data transfers at modem-level speeds.

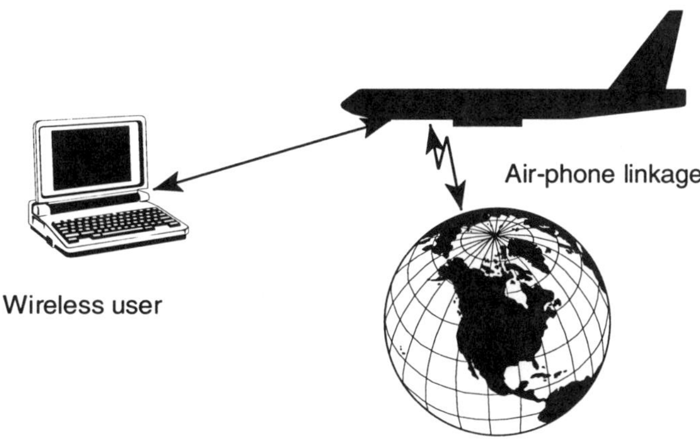

Figure 12.13: Use of Air Phone- to-LAN Connection

Wrist communicators

As the wireless communicators continue to decrease in size, they may become the realization of the famous Dick Tracy wrist communicator. With the ability to store information and connect via wireless technologies to the rest of the communications networks, the wrist connection would be convenient and serviceable for many users.

The current product that comes closest to this long-held future vision is the Timex data watch. By using software on a personal computer, you can develop a schedule and a database and then download it to the Timex watch by placing the watch in front of the CRT screen with the data and the supplied program active. The data are stored in the watch and can be recalled to the dial face via a set of push button commands. At $99 you get a multfunction watch and a wrist computer communicator.

CHAPTER 13

Wireless LAN Network Management

Wireless LANs will need to be integrated into the overall communications network management scheme in use by the organization. There are several parts that will be unique in the management of the wireless LAN world, but the overall management is only a subpart of the total communications operations.

Wireless LAN management will focus on the local user issues and the technology integration points between wireless and wired service worlds. The overall end-to-end management of data and operations will be part of the total network management layer.

This chapter will focus primarily on the issues that are unique to the wireless LANs. It will touch on but avoid detailing the operational aspects of total communications systems and the management of the wired worlds. It will show how transactions and information flows are handed off between network layers and how the initiating layer must maintain some overall control on the return and completion of the service process.

Service tracking

The wireless LAN is a user-oriented service layer. As such, it will be interfacing with users and taking and delivering communications results for the user. Most of this level will depend on the user initiating a request for service or information transfer. These issues of service requesting and service tracking are paramount to the management of the services at the wireless LAN level.

Service tracking requires that the network management system be able to identify, log, and trace the flows of information from source to destination while it is in the network and determine that the steps and results meet acceptable standards.

The service tracking process involves a mix of data tracking and service locating. Once a service is requested, some part of the wireless LAN will have to capture the identity of which network component has been involved in providing services to the data traffic. The service depends on using the identification information that is contained in the transmitted packets. The network component tracking involves using assigned identifiers and collected information maintained on the traffic flows at various points in the network.

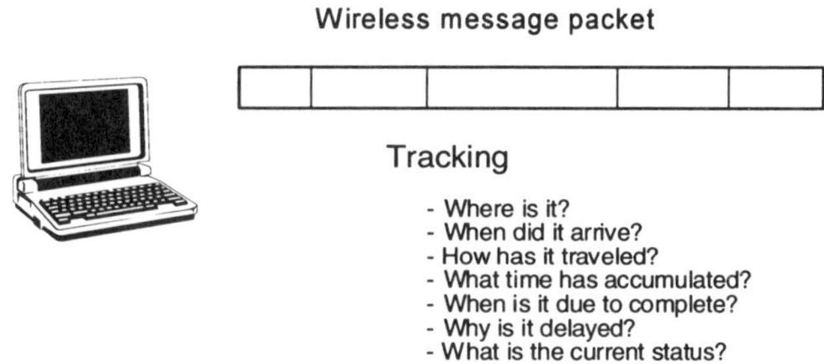

Figure 13.1: Tracking of Wireless LAN Services

Most elements of service tracking can take advantage of available standards for identifying the data and component elements in networks. The data identity will be provided by the data level of the protocol used (Ethernet, token ring, ARCnet, etc.) and the physical network processes defined by a hardware implementation of Simple Network Monitoring Protocol (SNMP) or Common Management Information Protocol (CMIP) or the Remote Monitoring (RMON) standards used in the network hardware levels. By combining these two elements in a network management and control level, the organization can monitor and manage the overall network and data flow processes.

Data identification

The first problem in managing a wireless LAN is to be able to identify what traffic is being handled and establish its source and destination. After that the validation that the data contents of the message are correct and error free should be tested.

The tracking of data in a wireless environment can be difficult. The data can be hopping across various frequencies in its travel between the wireless units. The data can also be stopping at different control and conversion points as it makes its way on an end-to-end journey.

An attempt to log all of the send-receive-retransmit points could be extensive, expensive, and produce more data than could be evaluated or analyzed. The best data tracking may be to organize test messages that are sent through the wireless world and evaluated at their send-receive points. This test process could be invoked as needed, or run on a background basis to validate that the system and its components are working correctly. The key is to be able to assure users that the wireless LAN is a reliable service that they do not have to worry about in terms of the information accuracy and delivery guarantees.

Coupler identification

The various units that will make up the wireless LAN world will all carry some form of identifying codes. The Ethernet units will conform to the 48-bit fixed address definition of the standard. All token ring couplers will carry assigned names.

In addition, many of the components of the wireless LAN world, such as servers, bridges, routers, and gateways, will carry capabilities to report their states and operations via SNMP or some form of RMON capability. These protocols will provide identification of the units and a way to interrogate their status and condition via the network and some standard protocol requests.

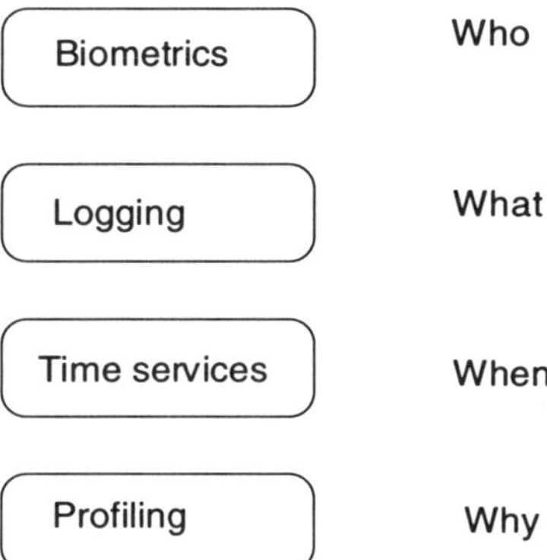

Figure 13.2: Using automatic identification concepts

The identification of the units will be defined in their header information file. This information will be passed through the network by the servicing protocol.

Probes

Probes are used to detect and report conditions at selected points within a network. By placing probes at strategic points within a network, the monitoring system can interact with the probes and request status and error information to be collected and transferred to the monitoring system. The probes are sometimes intelligent enough to raise an alarm message when a fault occurs. Other probes simply collect data and must be queried to upload their data for event analysis.

Probes on a wireless LAN network would usually be placed at the servers or at the hosting point where the wireless systems come together for servicing. This reduces the number of probes needed. As the wireless services will all flow through these probe points, it makes all of the wireless segments subject to probe evaluation and support.

Monitors

Monitors, as in display monitors, are used to present and evaluate in real time the status of a communications network. Monitors would display the structure of the network and the traffic flowing through it. The monitors can be shifted to any probe point and can zoom into varying levels of detail on the monitored performance of the network.

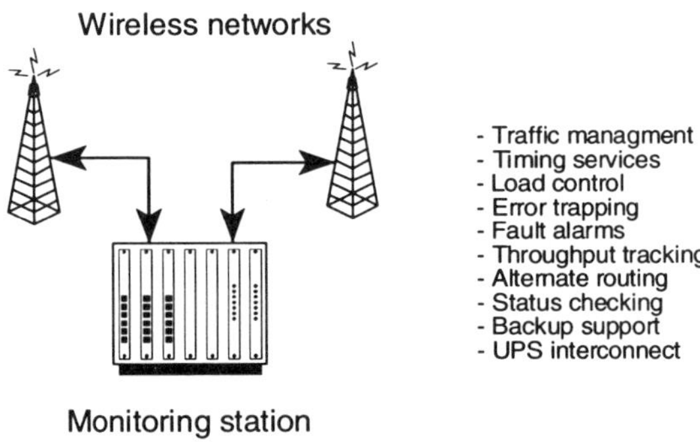

Figure 13.3: Monitoring a wireless LAN world

Wireless networks need to be monitored as to the state of the wireless devices and the level of usage traffic flowing over the wireless connections. The monitor data are collected from probe points or from logical collection points where individual and collective traffic can be identified and quantified.

Traces

Traces are specific requests to follow the steps and transfers in a stream of communications activities and show the end-to-end movements and timings as the transmission traveled through the network. In the event of an error condition, a trace can show how the transmission traveled and where it became flawed. Some traces are historic; others are newly created messages that will follow a predefined route and keep track of the detailed happenings at each point.

Loads

The monitoring of a wireless LAN should be able to define the loads that it is carrying and the source and destination of the loads. With this information, the systems manager can determine if the performance is within reason and if the wireless segments are providing reasonable service to the users.

Given the lower speeds of the wireless LANs, load variations can have a greater impact on throughputs, errors, and performance than in the wired LAN world. If the transactions are kept short and the return data deliveries are tightly packaged, then the load levels should be appropriate for the wireless world. If the levels, sizes, or demand levels become too large, then the wireless LAN will fail to deliver adequate performance to the users.

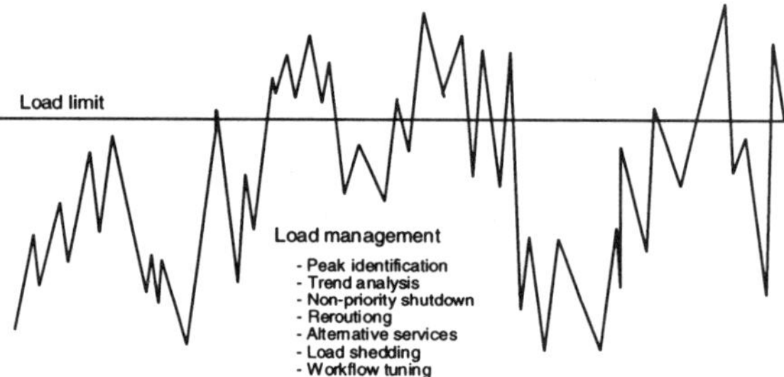

Figure 13.4: Load management in a wireless LAN

By tracking the load levels, the systems managers can identify trends that will lead to less than satisfying performance and take steps to reduce or eliminate their adverse impacts. Load information should be collected at all wireless-to-wired world entry points. The levels of both directions of traffic flow as well as some information on the sizes of the overall data flows should be recorded.

Audit trails

Who did what, to whom, and when is the recording responsibility of an audit trail. Wireless LANs may from time to time need to produce an audit trail to determine the answers to these questions when some type of failure has occurred. Audit trails should be selectable for specific locations and times within the wireless LAN architecture. To run network audit trails on a regular basis would generate too great a volume of captured information. Selected running between defined probe points will provide sufficient details to define the flow and characteristics of the traffic without consuming an excess amount of storage area.

Once the audit trail has provided the information or confirmation of the message movements, it should be turned off and the stored data purged. Unlike financial audit trails, which are archived as records of actions and fact, the network audit trails are used only as diagnostic tools.

Reports

The probes within various locations of the wireless LAN network can be used to sense the data traffic flows and generate statistics that can be used for reporting on overall loads, throughputs, and performance. The reports will be support for the assessment of the overall health and performance of the wireless LANs.

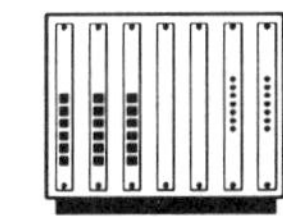

Reporting outputs

- Usage statistics
- Access logs
- Traffic flows
- Volume statistics
- Load levels
- Response times
- Message sizes
- Faults
- Outages
- Delivery times
- Packet strings
- Time of use
- Calendar variations

Figure 13.5: Reporting on Wireless LANs

Historical analysis

Over time, a history of performance should be developed for the wireless LAN. The history should cover the response time trends, average loads, performance factors, problems, faults and errors, and the overall operation of the wireless LAN. History not only repeats, but it can be used to determine factors such as aging, excessive failure rates, overloads, business load factors, and other time-and-condition-variable situations.

Faults

Fault management and control is a major part of wireless LAN management. If the communication link is down or goes down while in use, it must be detected and corrective actions taken quickly and judiciously. It is easy to perform cable tests on a wired system; on a wireless system is far more difficult to do fault detection. The air medium seldom fails, but interference from the environment or failure of the transceivers is a possible cause of faults.

Self-testing is one way to assure that units are functioning. Another is to send background communications packets to test that all is well within the system. Both of these techniques would have to report their findings to a higher level of management system in order to secure the necessary attention to address and repair the fault.

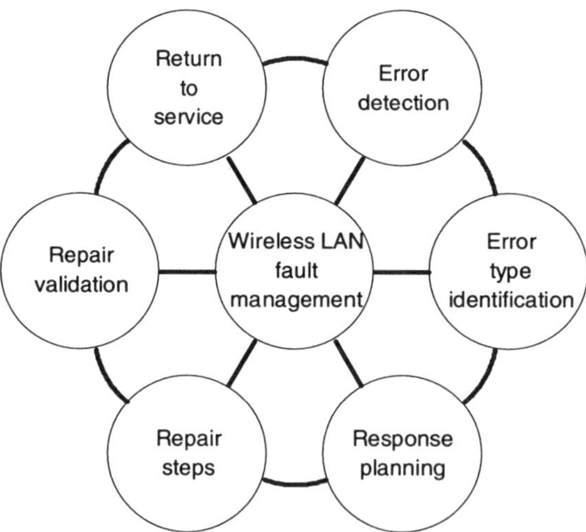

Figure 13.6: Wireless LAN fault management

Losses

Loss of a message or a piece of a wireless data communication is a very likely and high-probability occurrence. By itself the loss is a concern. If the system can detect and correct the loss quickly and efficiently, the problem can be minimized. However, if, as in the cellular communication world, some communications are just cut off and abandoned with no sense of management or continuity responsibility, the wireless LAN services will be declared unusable.

Losses of messages in whole or part is a critical concern for wireless LANs. The solutions lie in the detailed message-by-message management of the end-to-end delivery of communications.

Errors

The first step in reliable transmission is to know when an error has occurred. If there is no indication of an error, the erroneous data will be passed to the user and the error will become part of their business process. If, however, the error is trapped and identified, the system can remove the bad data and take corrective actions. But if the error passes without detection, the operation of the wireless LAN is compromised.

Incomplete transmissions

Some errors are partial. The wireless LAN sends and receives part of the message. The part received is good, but it is only part of the total message. Incomplete transmission means that a part of the message has been lost. Although the received good part of the message may be useful, it is not the total message transmitted and is thus there is an error of omission.

Incomplete transmissions are often very difficult to trace and correct. There is no error within the delivered message. In addition, the break in the message may be random, making it very difficult to diagnose. Audit trailing of all messages may be needed until some pattern of incomplete message can be discerned.

Lock-ups

When a system locks up, everyone knows it. The good news is that the error and the fault are readily defined. However, the fix and corrections are more difficult to define and implement. Lock-ups can occur at any time and often for no discernible reason. They can range from power blips to interference in the network.

Lock-ups in wireless LAN networks will most likely mean the user cannot access the network or any of its services. A lock-up can occur at any time, so a user may be in the midst of a transmission and have the system come to a lock-up halt. In other situations, the user is inactive and cannot access the wireless network when trying to enter a transmission mode.

Lock-ups are especially frustrating to users, because they do not know what has happened, only that the system does not work. The fault is more visible, but the cause and fixes are more hidden.

Disappearances

When a message disappears from the wireless LAN, it disappears into the science fiction ether. A message can be in process at one instant and then gone the next. Where it went and why can remain a mystery.

Disappearances are losses, but they are complete and mysterious. The only answer is to retransmit the entire communication. Fortunately, most disappearances are random occurrences. This means that there is a high chance that the second transmission will be completed successfully.

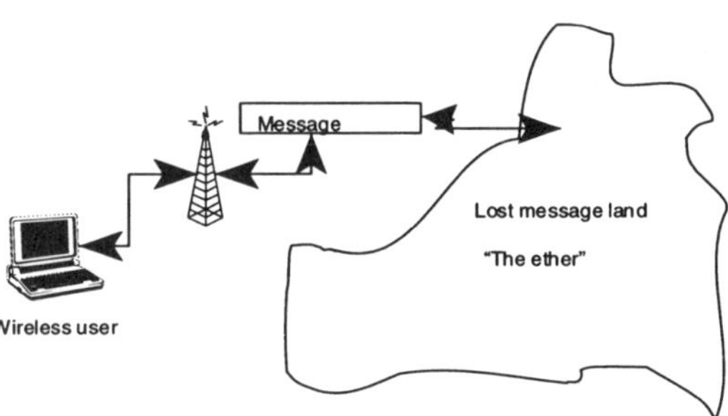

Figure 13.7: Message disappearance on a wireless LAN

Disappearances are a frustrating part of wireless LAN management. They leave no audit trail and seldom return. They are like the noise in your car that always disappears when you take it to the service station. The cause of disappearances is usually some slightly out-of-balance condition which goes away and refuses to stay in place. The good news is that the disappearances often disappear. The bad news is that they choose to reappear at random times and locations. Maybe there are technology ghosts that haunt wireless LANs.

False packets

When a local area network sends packets that are not user or system initiated and do not contain legitimate data, it is creating and sending false packets. These packets take up space and consume available bandwidth. They can overload the system and freeze legitimate users out of their needed services.

False packets are often generated as a broadcast storm from a broken network interface coupling (NIC). The usual solution is to locate the source of the false packets and then shut it down and correct the problem. False packets can be detected and tracked. They can be picked up by remote probes and used to automatically isolate and limit the extent of the false traffic on the total network.

Virus infections

Virus infections can be introduced by the mobile users of a wireless LAN when the independent units become contaminated and then pass the virus on to other stations when they connect via the wireless LAN. Virus infections can easily pass the probe monitoring points in the network, because the probes are looking for network errors and the virus is implanted within legitimate data packages.

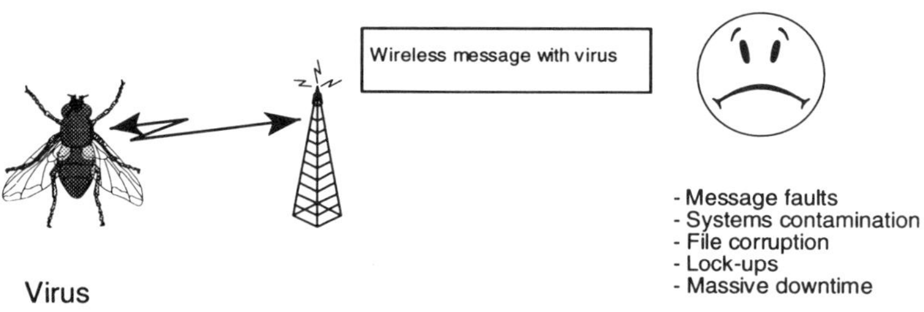

Figure 13.8: Wireless LAN virus infection

The major way to trap wireless LAN viruses will be to scan for viruses all transmissions going from the mobile world into the server. If any suspicious data are seen, the virus checker should either halt the communications process or remove the virus component and its damaging effects.

Fraud

When users conduct their business activities to perform an illegal or self-serving resource action with a data/information process, they may be committing fraud on their organization. Fraud can certainly occur within wireless LAN systems. However, the use of probes and test points will not be likely to capture the existence or processing of fraudulent operations.

Fraud must be captured by data value inspection, financial audit trails and management oversight systems. Wireless LAN network monitoring and controls are not organized or intended for use in the monitoring and prevention of fraud.

Security breaches

Security breaches are probably the most worrisome errors in a wireless LAN environment. Some of these concerns are based on the fear that air-based transmissions can be easily subverted, copied, or modified. In actual fact, the hopping of the radio frequency LANs and the tuned relationship of infrared and microwave severely restrict the opportunities for undetected security violations in a wireless LAN. In fact, it may be easier to breach a wired LAN than it is a wireless LAN. The wire is easily located and copper can be capitance/inductance monitored without detection to copy/modify messages within the network.

However, if a security breach does occur in a wireless LAN, it is just as significant as one in a wired environment. Fortunately, this is less likely to occur, but when it does, it may be far more difficult to locate and correct. Any security violator would know of the wireless area coverage and be able to connect and disconnect their wireless LAN violations quickly and with minimal chance of detection. This means that continuous monitoring of users, uses, and messages within the wireless LAN world is the only reasonable protection from someone breaching the security of the network and causing some damage to the messages and content of the LAN.

Fault monitoring

The operation of a successful wireless LAN environment requires that the wireless services operate at a high level of reliability and be available to the users when they need them. Any failure to provide reliable service between the users and the wireless nodes would be deemed a problem and put the success of the applications at risk. The correct way to handle possible faults and problems in the wireless world is to set up a proactive operations procedure of monitoring transmissions and isolating and responding to faults and problems.

Fault monitoring can be done in background with a small amount of overhead in the transmission process. When faults are detected, automated correction steps can be initiated and the errors corrected or the messages retransmitted. If the errors are small and intermittent, the fault monitoring and response process will appear transparent to the end users. If the faults are more substantial, causing outage of network capabilities, then the monitoring procedure should seek to provide alternate services and inform the user of the situation. It should also initiate actions to repair and validate the correct return of the wireless LAN capabilities. Notices to users on all states and conditions should be automatically generated and transmitted.

Occurrences

The fault monitoring process will need to be continuously vigilant to the operations of the wireless LAN. When a fault occurs, the monitoring system should trap the error and initiate an automated response process. The occurrence should be reviewed to determine if it is a sequence of faults that are recurring or if the condition is a random, one-time occurrence.

Repeat occurrences need a high-priority attention. The first occurrence of a random event needs correction but not necessarily a priority response. The occurrence should be logged, noted, and evaluated for prior existence. The more significant the occurrence and the more repetitive, the more significant the corrective response.

Events

The event is the actual occurrence of a fault within the wireless LAN world. It has the identification of what occurred; when and how it occurred, and how it was detected. It can also contain a picture of what else was happening at the time of the fault. The event should be logged, identified, and used as the data reference component.

Isolation

Isolation is the process of providing a picture of the fault event and the conditions under which it occurred. The isolation steps would provide a snapshot of the fault event and the environment in which it occurred. Isolation data can include all of the rest of the world that existed when the fault event occurred. Obviously, this is too much data to capture and process. Most isolation data involve capturing the state of key variables and the details of some of the surrounding loads and traffic conditions. If this does not provide sufficient data, then further diagnostic monitoring will need to be initiated to trap other relevant information.

Recognition

Once a wireless LAN fault event is identified and isolated, it must be classified in terms of what it means and a decision made on how to respond to the event. Recognition would attempt to type-cast the event and develop a rational (and automated) response to the occurrence. The recognition process should evaluate the fault event in terms of whether it has been seen before, its importance to the wireless LAN operation, and any stored prior response steps. If recognition does not find a match, then other steps, possibly including operator intervention, would be needed to resolve the next steps in the error management process.

Corrective initiation

Once the event has been defined and recognized, it will lead to a decision process on how to deal with the corrections and cures for the event. If the event is a random occurrence, there may be no immediate corrective response. If the error involves user data, then the first response will be to retransmit and test the results. If the error is seen to persist, then the operations procedure should move to active corrective initiation and turn the fault and event data over the fault correction process.

Fault correction

Fault correction is the process of responding to a fault that is significant enough or reoccurring to the point that a measured response is needed. The fault correction process will determine the severity of the fault and the priority of response needed. The process will also manage the response and oversee the actions taken to effect corrections to the problems. Logging of the conditions, responses, and results will also be done to build and maintain a history that can expedite future responses to similar conditions.

Typing of the fault

Before initiating actions in fault correction, the process has to determine what type of error has occurred and the level of response that is needed for the given situation. Typing of the fault involves identifying the occurrence and the conditions under which it occurred and trying to determine the overall significance of the error event. Typing can be done against a stored library of previously classified faults or against some sort of scalar rating scheme covering the various types of faults.

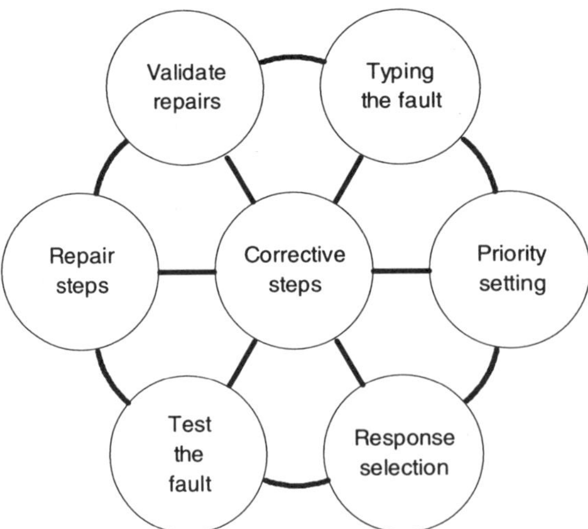

Figure 13.9: Corrective Steps

The typing alternatives could be defined by the options presented in the following list:

- Network interface error

- Media transmission error

- Protocol fault

- Lost message

- Incomplete message

- Interference

- Noisy media

- Unknown error

- Server response error

- Hardware fault

Priority setting and response selection

After the wireless LAN error is defined as to type, an evaluation of its severity and priority of response should be defined. If the wireless network is totally inoperable, the severity is critical and the response priority should be high. If the problem is sporadic and does not prevent retransmissions from providing some level of valid throughput, the error would be determined to be of modest severity and the response priority set to important but not critical. Urgent may be a proper definition for this scenario. If the problem is random and does not reduce the performance of the overall system, the severity is minimal and the response priority should be to increase the vigilance.

Priority setting can be descriptive or ranked numeric. Whatever scale is used, the priority setting should be used to engage and manage the response resources as they attempt to refine and deal with the fault condition.

Testing the fault

Once the fault has been typed and prioritized and a response defined, it is useful to test the fault to determine if it can be reproduced and if the conditions which were existent at the time of occurrence have changed. If the fault cannot be reproduced, it may be a random error that may never occur again. This error can be logged and forgotten. If the error is repeated, then the response plan should be implemented.

The fault test should be quick and cursory. If more detailed tests are to be made they should be part of the solution analysis and corrective action process.

Repair steps

A wireless LAN fault is typed, prioritized, tested and validated. At this time the systems manager should be able to determine the response actions and repair steps to be initiated and performed to correct the condition and return the system to reliable operation. The repair steps may involve the following:

- Further investigation of the fault and the conditions under which it was observed.

- Comparison of the fault to the fault library to determine prior occurrences and responses.

- Identification of what and who is responsible for the response.

- If the cause is laid to an external party, then call them and transfer the responsibility.

- Invoke the repair logic best suited to the situation.

- Establish mechanisms to test the repairs for correction of the fault.

- Implement and test the repairs.

Validate the correctness of the repairs

Any change made to parts of a wireless LAN world needs to be tested under realistic conditions to determine that the repair corrects the fault and does not introduce failures in other areas. This is done by running a suite of predefined test cases that will test the overall operation of the total system. In addition, specific tests may need to be added to the suite to test the adequacy of the specific repair made to correct the reported fault. This process may require several tests to convince the fault response team that their changes are stable.

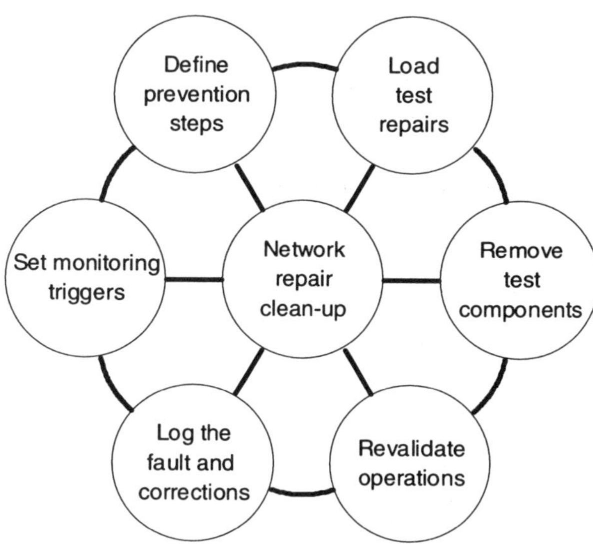

Figure 13.10: Clean-up Steps

Removing tests-clean up

After the repairs have been tested and validated, the systems environment should be cleaned up and returned to its operational state. This involves the removal of any testing components, logs, test points, probes, or other components used for identifying and correcting the fault. After these are removed the test suite should be rerun to determine that no faults have been introduced by the removal process.

Logging the fault and correction

After the fault is corrected and the system returned to operational status, the fault repair team should document their work and record suitable entries into the fault history library. If any steps in the repair process have changed, they should be noted as corrections and/or improvements to the repair procedures for the fault. If the repair is a new one, then full documentation of the steps taken and the results seen should be prepared and recorded. Any indices or cross-reference entries should also be updated.

Watching for success

It is a good idea to set up an occasional monitoring window on the recently completed fault and its repair to test if all is holding together. If the repair fails, it will likely be in the short term. By monitoring the conditions of the fault for recurrence, the repair team can convince themselves of the success or lack thereof of their work. If the fault does not reoccur after a period of time, the monitoring can cease and the system return to normal operating mode. If the fault reoccurs in any way, the repair team will have a rapid identification of the occurrence and can move quickly into the evaluation and repair process with minimal front-end analysis and decision making.

Documenting for the future

The final step in the fault recognition and repair process is to document the findings and knowledge gained in the form of information that can be reused in the future. This could include recommendations for future repair steps or ways to monitor the system for earlier detection of the fault.

The key to this documentation is to find ways of improving the systems monitoring process and to make the overall system better and more stable. The intent is to reduce error occurrence and make the system operate more reliably. Any advice that can make such improvements is worthwhile. The documentation should also provide any advice on ways to improve the fault finding and correction process. This would also generate valuable improvements to the system and its overall quality management.

Error management

Errors are generally more likely to occur in a wireless LAN than in a wired LAN environment. This is due to the signal quality of the air medium, the contention for the limited signal bandwidth space, possibility of interference, and numerous other factors. The above steps have shown how to detect and respond to reported faults in the wireless LAN world. Error management will be the overall process of controlling the fault repairs and evaluating the overall system for its performance and error history.

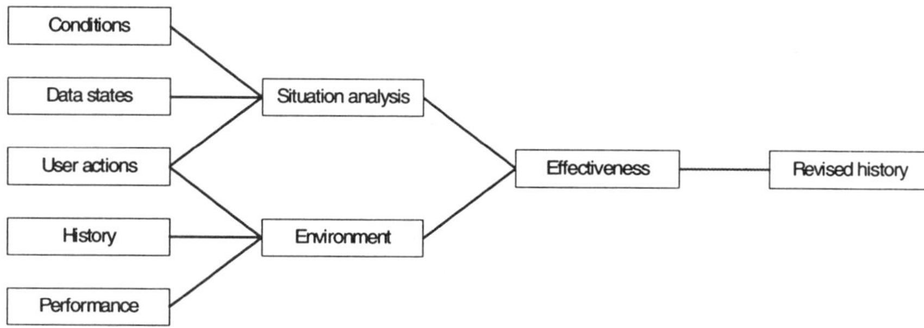

Figure 13.11: Error Tracking

Error management will provide the record of how well the system has performed and the effectiveness of the fault response and repair practices. It should develop a set of performance measures and use them to evaluate and report on the overall state of the wireless LAN operation.

Error Reporting

The error management process should evaluate the actual history of loads, errors, faults, repairs, outage periods, and other problems. These conditions should be addressed in a set of reports that provide the telecommunications network management unit with a comprehensive picture of the state and operation of the wireless LAN segments. The error reports should provide a history of the operation of the wireless LAN and supply a basis for making decisions on changes, improvements, or adjustments to the wireless LAN use within the overall organizational network strategy.

Performance Improvement

By studying the error histories and performance logs of the wireless LANs, it should be possible for the telecommunications experts to determine recommendations for improving the state and performance of these units. Improvements may require hardware changes, procedural adjustments, or other modifications. The key is to use the actual results of errors, problems, and loads to define ways to make the system and its services to the user community more reliable and stable.

Upgrade management

Wireless LANs will be in a state of continuous change and improvement. This means that there will be an ongoing evolution of upgrading and replacement of the technology levels. This replacement is typical of new and fast-growing technologies. It is an expensive proposition that must be carefully managed to assure that the organization receives reasonable value for the efforts invested.

Upgrade management should strive for maintaining some consistency in the number of levels of different technology generations and products supported via wireless LANs.

Limiting to three or four products operating at a limited number of standard levels would be the most desirable. This will minimize the diversity and support requirements and allow the organization to leverage its volumes during an update of the technologies.

Changing technologies

The technologies of wireless LANs will be changing over the next several years. There will be an active search for more reliable and faster ways to couple devices to LANs without wires. Organizations will need to face the fact that the current technology will likely be obsolete before the working life ends with current technologies. Wireless LANs will be similar to PCs, which have been in a constant state of turnover since their introduction.

Changing components

The main components that are likely to change in wireless LANs will be the transceivers and the couplings to the wired LAN world. These components are the peripherals of the wireless LAN. The key change will be in favor of improved speed and reliability. The costs of these units are also likely to decrease, making the wireless LAN systems more practical and affordable.

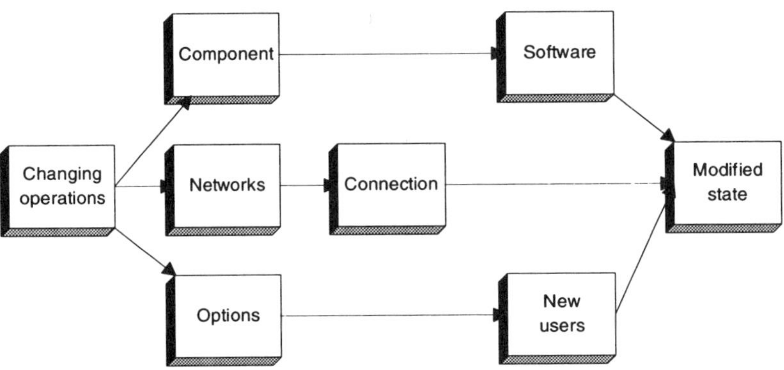

Figure 13.12: Growth Management

Changing connections

The connection or connectivity point where the wireless and wired worlds meet may also change in the future. The current point of common intersection is usually the shared server. This may change to the intelligent switching hub. Such a shift would provide the wireless LAN with entry into any part of the wired LAN world, rather than just the server levels.

Changing options

The changing of options includes adding new features and facilities for the end users. Such features could include integrating voice calling with the wireless LAN such as in the IBM Simon product, providing wireless electronic mail service, connectivity to the Internet, and a host of other user services.

When options are changed, the user population will have to be informed and re-trained as needed. Very few new options will be automatically interpreted by the populace. Most will need explanation, examples, and some level of user training.

Changing software and applications

The changing of software within a wireless LAN world will probably be the most continuous change. New software products for the mobile users and upgrades to existing applications will be a constant stream of potential change. Not all new software nor even upgrades to existing software need to be implemented at the operational level. New products should be thoroughly tested and evaluated before adding them to the operational complement. Even new upgrades to existing applications should be reviewed to see if the changes are useful and needed by the organization before moving to a wholesale upgrade.

Software changes will occur and will likely be part of an ongoing evolution for the mobile users. These changes can be unsettling to the smooth operation of the system. By taking some extra steps, the software change process can be improved for the wireless LAN users. The extra steps include:

- Prenotification of the pending change

- Demo trials of the changes

- Early training on the new processes and operations

- Availability of HELP support

- Continuous contact and communication on the acceptance and understanding of the changes

- Use of the network for help and support

Wireless LAN users are often more distant from other users as they operate in their mobile world. This makes the implementation of software changes a more person-dependent process with less direct help and support through the learning and transition time. With support and communications provided through the wireless LAN services, the users will feel more comfortable with the change process and the move to the new or changed applications.

Adding new users

The addition of new users to a wireless LAN network should be very simple. They are given their mobile connection client, trained how to use it, and then turned loose to do their new job. However, as wireless LAN users are more separated from the rest of their operating world and usually operating independently, they may need more time and support to become fully conversant and comfortable with the wireless system.

The following steps could be taken to assist new users through their initial transition period with a wireless LAN system. Provide:

- A big-brother or big-sister connection

- Easy-to-use on-line help

- Direct wireless communications of questions or problems

- On-line examples of how to do major functions

- Walk-in areas where help is available

- Regular contacts from knowledgeable users

- Welcome activities from other wireless LAN users

The key to adding new users to a wireless LAN world is to make them feel welcome and keep them from getting lost and frustrated by the new technology and independent operations. Once they are over the transition period, the new users can be used to welcome other new users.

CHAPTER 14

The Future Evolution of Wireless LANs

Wireless LANs are still in their infancy. They will move through several generations of new and improved technologies on their way to maturity. The evolution for the future of wireless LAN's will include:

- Higher speeds

- Improved security

- Multiple frequency hopping

- New vendors

- Seamless end-to-end protocols

- Better error control

- Longer distances

- Home systems

- New devices

Vendor movements

Wireless LAN vendors have been composed of a narrow set of technology innovators who see the wireless world as a key part of their future strategies. There are big and small players in this vendor field. Some names like Solitron are small firms whose total business is wireless LANs. Other firms such as AT&T, Motorola, and IBM see wireless as a key component of their enterprise networking approach, but it is (and will likely remain) only a small part of their overall revenue and business stream.

The movement in the vendor arena is still in the establishment of the wireless LAN marketplace and the seeding of the concept to the buyers and planners of LANs and network services for the future. In addition, the vendors are trying to define their specific choice of technology for wireless LANs and focus on the applications they believe are best suited for their technologies and support. Some of the current vendor choices in this area include:

- AT&T Global Information Systems spread-spectrum technology

- IBM infrared

- Hewlett-Packard infrared

- Motorola microwaves

Most of the smaller vendors have also selected specific technology or application niches for their concentration in the wireless LAN world.

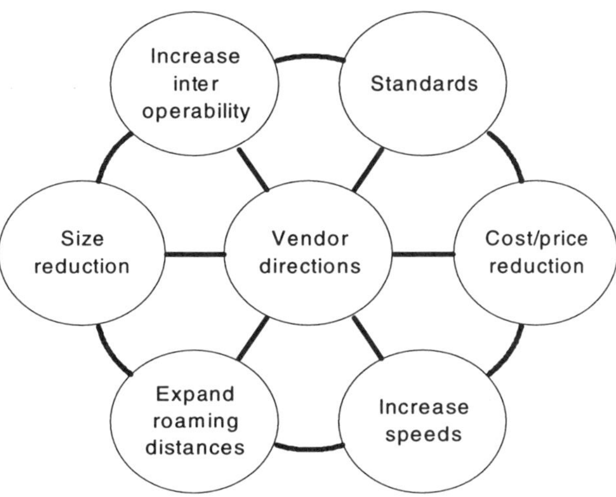

Figure 14.1: Vendor Movements and Directions

Standards development and acceptance

The standards for wireless LANs will be shaped by several sources. The major standards will continue to flow from the work of the Institute of Electrical and Electronics Engineers (IEEE) under their work on the 802 series, which fits within the ISO/OSI umbrella of standards. As the IEEE was given the task of developing LAN standards in the late 1970s they have a logical base of including the newer moves to wireless types of LANs.

Unfortunately (or fortunately), the IEEE official standards work is slow and plodding. It also tries to stay to a conservative middle-of-the road approach. Some technologies move much faster than an official giant of a standards body. This makes it difficult to develop standards while the technology is still evolving. The working alternative is that interested parties will often band together to form special interest groups (SIGs) which can focus on the key issues of the technology and develop recommendations for standards that can become de facto standards or be submitted to the official standards writing organizations as "draft" standards. These special interest groups are often called "forums."

The wireless standards arena has the following groups as active participants in the early formulation and drafting of standards proposals:

•   Infrared Forum

•   Wireless LAN Association

•   Microwave Industry Association

•   Personal Computer Memory Card International Association (PCMCIA)

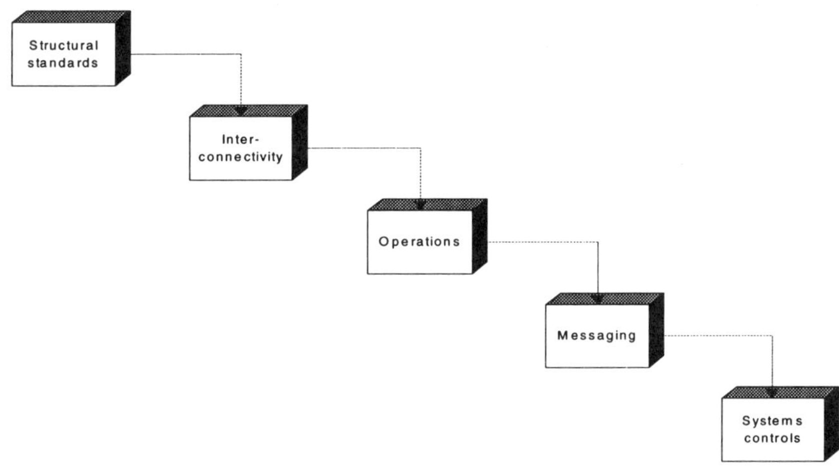

Figure 14.2: Wireless Standards Activities

Integration with the total networking world

The wireless LANs world is unlikely to become a major stand-alone world of its own. Wireless LANs are a part of the larger interconnection of businesses and enterprises to perform their activities in a more timely and productive manner. This means that wireless LANs need to establish a position as a player in a larger marketplace. Their role is to provide their unique mobility capabilities to users and systems that need these types of services.

Most wireless LANs will be the downline user connection and communication services to the rest of the organization's information world. The connection of wireless to wired networks will be a critical coupling that will expand and enhance the services and uses of the wireless systems.

The integration of wireless LANs with the total networking world will be through standards and software that permit the wireless LAN to be a seamless extension of the information services of the organization. The wireless LAN will take its place alongside the end-user client services on desktop systems and the interfacing to the server-level architecture of the organization. Other wireless LANs such as the air bridges between buildings will be imbedded within the network architecture and will be totally unseen and invisible to the end user. Their services will be simply a dependable part of the overall network services world.

Wireless LANs may be free-standing, fully wireless systems with their own servers and resources. These systems would serve isolated locations or very small business units. Most wireless technologies will be used as a link to the more complete enterprise network. They will fill voids where other technologies cannot provide the services as efficiently or cost-effectively as the wireless LAN.

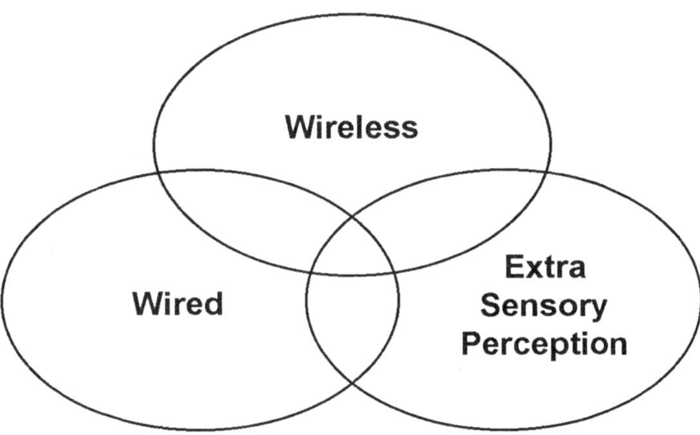

Figure 14.3: The Integrated Networking World

An aspect of total world coverage for the wireless LAN would be to extend it to the home environment. This would produce a wireless HAN (home area network) that would operate within the home environment and provide services such as:

- Data collection

- Energy management

- Security control

- Entertainment distribution

- Communication services

- Message centers

- Teleconferencing

- Personal information systems

- Data recorders

- Money managers

- Family coordinators

- Educational delivery systems

The wireless HAN would operate in a similar manner to an in-home portable telephone unit, except that it would provide for a broader and more complete range of services. Once HANs are popular, users could move from one environment to another and expect to have the wireless technologies adjust to support them wherever they went. For example, users can access their data system for messages from their home, access and create messages or transactions from their car via cellular wireless, and then at work switch to PCS services to couple to their information network as they move through their organization. This would produce total mobility and connectivity. It will need to be supported by a single user interface device that is able to switch automatically between the various networks and continue to offer consistent and reliable data services to the user. In addition, the same device may be used for the voice services and might be extended to carry video signals.

Enterprise strategies

Enterprise strategies for information systems and communication technologies should include wireless LANs as a subcomponent. The wireless LAN is not the major form of intercommunications for most organizations. However, it supplies a unique set of connective services that are appropriate for a certain type of worker and environment.

The wireless LAN deserves to be considered as an operational way to cover the local mobile workers. They can be a solid link in bringing a previously unconnected layer of workers into the corporate informations network. The wireless LANs can also be heavily engaged in supporting the active transaction work flow of the organization. As active data collection networks, the wireless LANs can provide connectivity services to blend transactions into the established corporate database and applications world.

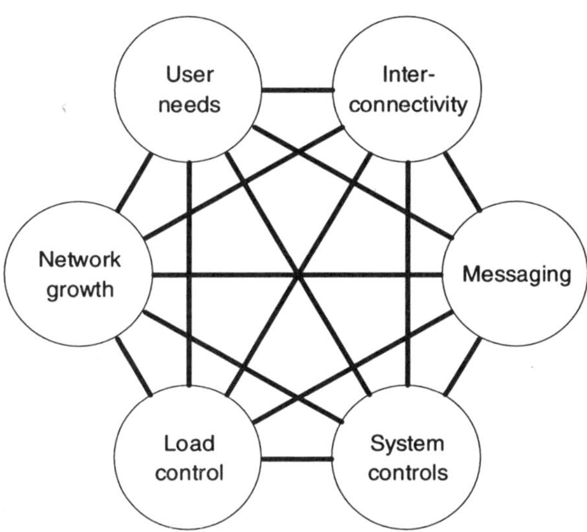

Figure 14.4: Wireless and the Enterprise Network Strategy

Global networking

Wireless LANs permit users and services to attain a higher level of mobility than with wired LANs. As wired LANs are being connected to the total enterprise network, so too will the wireless LANs have access to the same world of services. As the LANs become connected to metropolitan area networks (MANs) and to wide area networks (WANs) and eventually to global area networks (GANs), the wireless technology will receive access to the same world.

Global networking is the strategic direction of most organizations. As it happens, the wireless LANs will simply be a playing partner in the global access world. There will be a few additional restrictions applied to the wireless world in terms of operational management restrictions, but these are design and management factors, not limits of the wireless technology.

The limits applied to wireless technology in global networks will likely be in the area of access limits and extra security provisions. As it is more difficult to identify wireless systems and therefore they are more susceptible to fraud or unauthorized intrusions, the access from wireless systems may be limited for firewall protection and the entering users will likely have to pass through more levels of screening and identity validation. In addition, entries from the wireless world may be more restricted in terms of their access to data and services from the global world.

To some extent the wireless world has a stronger connotation of being local and not totally global. However, as systems increasingly focus on global activities, the wireless LAN can play in the same league if it is allowed to.

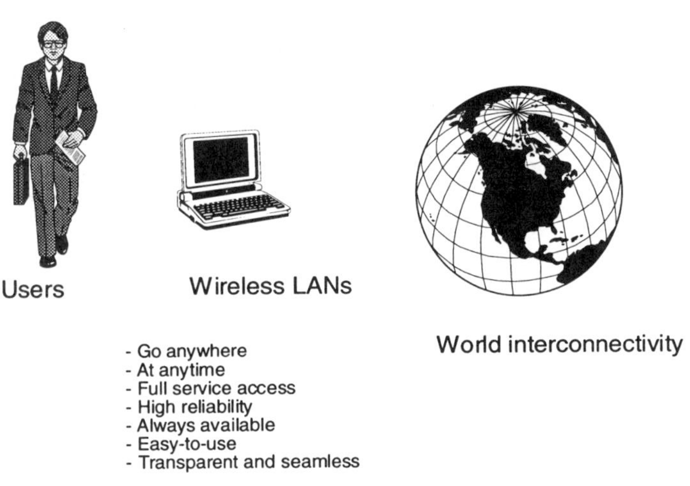

Figure 14.5: Global Internetworking

Advent of new technologies

New technologies are arriving all the time in the information systems world. Several new technologies will likely increase the popularity of wireless LANs. The most significant near-term arrival is the personal digital assistant (PDA). These small hanheld, portable units provide an excellent client interface for mobile workers. In addition to their local services, PDAs will need to interconnect to the larged enterprise data world. This interconnection will have to be via wireless LAN technologies, in order to provide the benefits of mobility and information access to the user.

The growth of PDAs is forecast to reach the hundreds of millions in a few short years. The rocky start-up phase is almost over, with stable technology, adequate power, and acceptable pricing becoming available in the marketplace. As this technology grows in popularity and acceptance, it will spur additional loads and growth throughout the wireless communications layer. Wireless LANs will likely increase their importance as the first entry layer to organizational services and access to the enterprise information world.

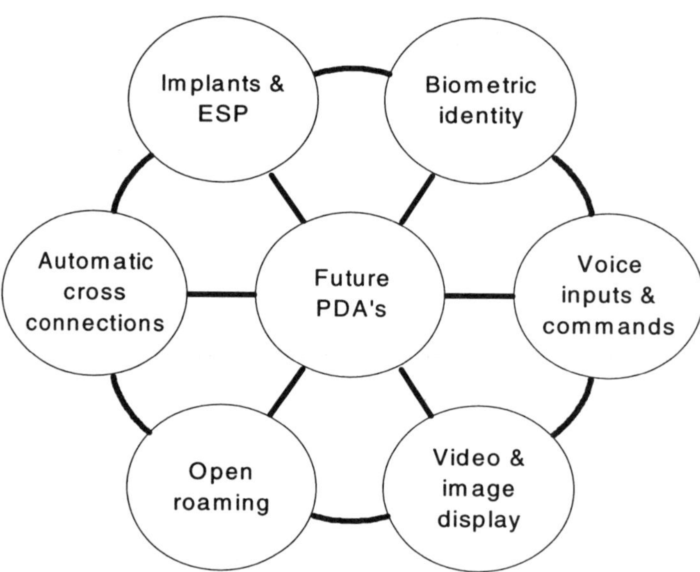

Figure 14.6: The PDA Revolution

Anywhere, anytime, any-form connectivity

The eventual level of interoperability will be one on which anyone can obtain access to the information needed for their work efforts, at any location, during any time period, and in any desired form or format. The wireless systems provide flexible support for expanding the location independence for information services access and processing. The wireless capabilities increase the flexibility of location for information actions, making many systems much more valuable and compatible with the information needs of the organization.

Seamless integration

The key to successful wireless LANs is that they do not place any unusual or different operational conditions on the users than they would experience in a wired world. In essence, the wireless services should be transparent and unseen. To the user the connections from a user/client system should be seamless to the rest of their permitted access world. The fact that a wire or a wireless connection is involved should be neither apparent nor a concern to the end user.

Seamless integration is not always that easy to obtain and maintain. There are physical differences in the wireless technologies and the operations. However, these differences should be contained within the hardware and software operating levels and be kept from the users. The more transparent and seamless the wireless world can be, the more succesful will be the technology and the satisfaction of the end users.

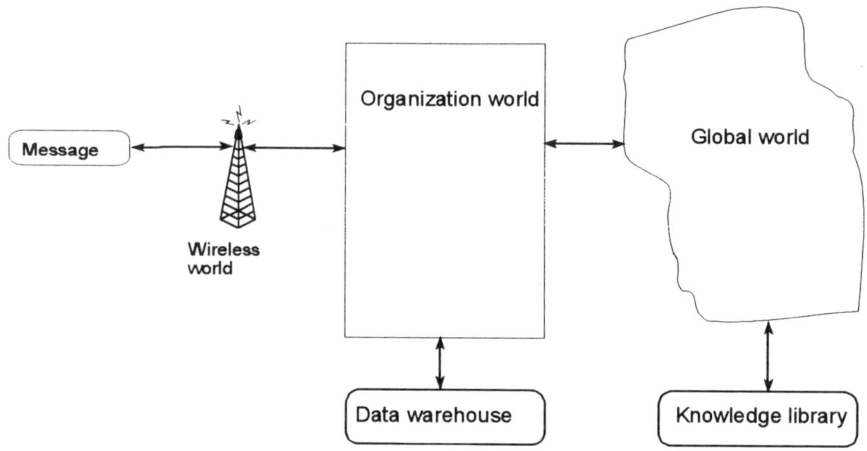

Figure 14.7: Seamless Integration of Messaging

Critical success factors

The critical success factors for wireless LANs include numerous factors of technology and usage within an organization. By evaluating and paying attention to these factors, an organization will be able to maximize its success in the implementation and use of the wireless LAN systems.

The critical success factors include:

- Suitability to the task

- Spatial and facility support

- Limits on user mobility

- Security of the information transferred

- Volume and speed containments

- User adaptability

- Control conditions

- Integration with other networks

If the critical success factors are positive, then the conditions are favorable for the implementation of a wireless LAN. This does not guarantee the overall success of the implementation; however, it means that with the proper steps and choices being taken in the design and implementation of the wireless LAN, the chances of success are good.

Index

E

F

G

H

I

T

U

V

W

X